Marine Electricity Handbook

Copyright © 2011
by
Marine Technical Training

Marine Technical Training
Academy

By: Alvaro Lopez
Edition 2023

Foreword

The book of electrical written by Alvaro Lopez is a well-defined skill that was written with many years of experience. This book serves to help the future Electricians, Mechanic, and anyone who is in any part of the marine industry who are in short supply. Not only is the book great for future technicians. It also serves as a reference manual if you forget any details in the electrical wiring. I am a well-seasoned Technician with over 20 years in the field and have worked with Mr. Lopez at the Marine Technical Training Academy for the last four years. I personally enjoy the information that the book provides as well as the videos. The videos are amazing as the book. It is almost like working along the side of an experienced technician when you are watching the videos. If you are an individual who has no knowledge whatsoever, I promise after you read the book and watch the videos. You will be starting to understand the electrical concept. That is the beauty of having this material at your disposal. The skill that you gain can never be taken from you. It is you for the rest of your life. Use it wisely and hope you enjoy it.

In conclusion, I hope everyone who reads the book and watches the videos enjoy it as much as I do. I would like to personally thank Professor Alvaro Lopez for his dedication to making these materials available for our future technicians four our marine industry.

By,
Daniel Rodriguez
Marine Master Technician & CEO Reel Loco Marine

Marine Technical Training Academy

Marine Technical Training is an online academy specialized in technical and vocational careers according to the USCG standards, ABYC standards, Federal Regulations, and NMEA standards for Marine Engineers, Naval Architects, and Marine Master Technicians.

This book is the recommended textbook material for the Marine Electricity course MTT 2420 of 160 Hours, four credits, and eight weeks of class.

Scan the following code to Enroll in this course.

Welcome To The Marine Electricity Course

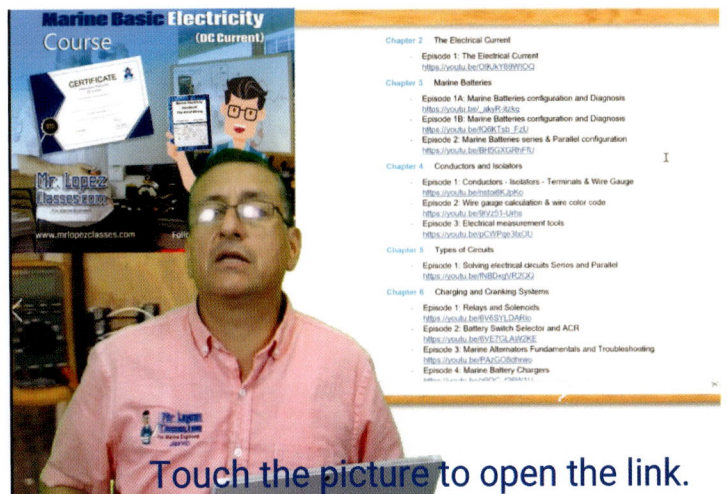

About the Marine Engineering Program.

This is a technical program focused on the training of Marine Master Technicians to cover the disciplines involved in the service, reparation, design and overhauling the pleasure vessels.

The courses are blended courses, 60 % on-line (Through of pre-recorded episodes of 45 min following the curriculum corresponding to the textbook of each course, divided in chapters), 20% in conferences and 20 % in remotely assisted labs.

The author of the videos and books is the Engineer **Alvaro Lopez** with 35 years of experience teaching in Mechanical Engineering faculties.

The material used in this program is property of **Marine Technical Training.** It is exclusive material with their own copy rights.

This program is composed by the following courses:

- -Intro to Marine Engineering — MTT 1004
- -Marine Basic Electricity — MTT 1400
- -Marine Advance Electricity — MTT 2420
- -Marine Electronics — MTT 2490
- -Marine Diesel Engines — MTT 2041
- -Marine Gasoline Engines — MTT 1073
- -Marine Transmissions — MTT 2234
- -Marine Generators — MTT 2042
- -Marine Auxiliary Systems — MTT 2541
- -Marine Air Conditioning Systems — MTT 1543
- -Marine Corrosion Systems — MTT 1062
- -Marine Composite Materials — MTT 1312

All rights reserved. No part of this work may be reproduced, stored or transmitted in any form or by any electronic or mechanical means, including information storage and retrieval systems, without the prior written and signing permission of **Marine Technical Training Academy.**

Manufactured in the United States By. **MTT**

ABYC Electrical Certification

This book was designed as a consult handbook for wiring, installation, and diagnosis of typical electrical issues in pleasure yachts. Also, can be used as a textbook for marine electrical courses and as a complementary study guide to take the **ABYC electrical certification.**

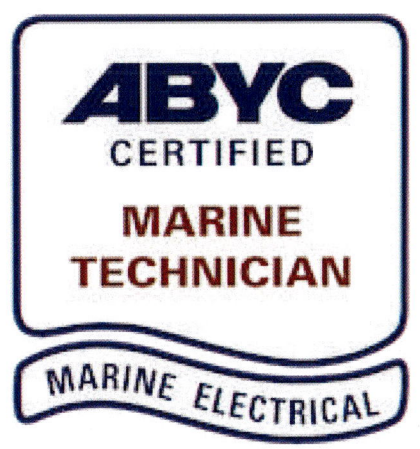

Table of Contents

Chapter 1 : The Nature of Matter	**9**
Atomic Number	11
Atomic Mass	11
Electronegativity	12
Ionization	13
Octet Ruler	13
Molecules	14
Oxidation and Corrosion	14
Static Electricity	15
Chapter 2 : The Electrical Current	**17**
The flow of charges	18
The Ohm Law	19
The Potential Difference	20
Types of Current	20
DC current	21
Chapter 3 : Marine Batteries	**23**
What is a Battery?	24
Marine Batteries Classification	25
Deep Cycle Batteries	26
Cranking Batteries	27
Bank of Batteries	28
The Electrolyte	29
Charging Process	29
Specific Gravity	30
The hydrometer	31
Lithium-Ion Batteries	32
Nickel-Cadmium Batteries	33
Amperes – Hour	33
State of Charge	33
Cold Cranking Amps	34
Battery load Tester	34
Storage Capacity	36
Battery Terminals	37
Battery Tips	38
Chapter 4 : Conductors & Isolator	**39**
Isolators	40
Conductors	40
Wire Connectors	43
Electrical Conductors	44
Circular Mills	44
Voltage Drop	45
Sizing AWG wire using ABYC Tables	47
Marine Wire Gauges	52
Marine Wire Color Code	53
Wire Sizes	55
Marine Duplex/Triplex wire	55
Continuity	58
Chapter 5 : Types of Circuits	**59**
DC Circuits	60
Open / Short Circuits	60
Series Circuits	61
Parallel Circuits	63
Combined Circuits	66
Power Calculation	71
Chapter 6 : Charging & Cranking Sys.	**74**
Types of Switches	75
Push Bottom Switch	76
Relays	76
Solenoids	77
Battery Switch Selector	79
Battery Isolator or ACR	81
The Alternator	83
Marine Battery chargers	87
Selecting the Charger	88
Cranking Systems	89
The Starter Motor	89
Starter Troubleshooting	91
Chapter 7 : DC Panel	**92**
DC Circuit Breakers	93
The Ammeter	95
The Shunt Ammeter	96
Reading Amperes	97
The Voltmeter	98
Chapter 8 : Marine Dashboard	**100**
The ignition Switch	101
The Fuel Gauge	102
The Coolant Temp Gauge	103
Types of Temperature Sensors	104
Troubleshooting Gauges	105
Analog and Digital Tachometers	106
Wiring the Dashboard in a Boat	107

Table of Contents

Chapter 9 : AC Current — 109
- AC Current Production — 110
- The Phase Angle — 112
- The Sine Wave — 113
- The Frequency — 114
- Frequency Vs RPM — 115
- American Power vs European Power — 117
- The Amplitude — 117
- Wavelength — 117

Chapter 10 : Bonding & Lightning — 118
- Bonding & Grounding — 119
- Cathodic Protection — 121
- Common Bonding Conductor — 122
- AC Ground Faults — 123
- Galvanic Isolator — 124
- Isolator transformer — 127
- Lightning Protection — 128
- Galvanic Corrosion — 132
- Electrolytic Corrosion — 133

Chapter 11 : AC Wiring — 135
- 50 Hz or 60 Hz — 136
- AC Wiring — 139
- AC Wiring Color Code — 141
- Types of Marine Wire — 143
- Neutral Connection — 145
- Grounding conductor — 147
- Types of Outlets — 148

Chapter 12 : AC Panel Board — 149
- Ship / Shore Switch — 150
- Triple breaker Switch — 151
- Double Phase Wiring — 152
- Double Phase Open Neutral — 153
- Double Shore cord Inlet — 153
- Two Shore Cords/Two Generators — 154
- Reverse Polarity Indicator — 155
- AC Wiring Warning — 159
- E.L.C.I — 161
- Faulty Ground — 161
- Shore Power Cord — 162
- Shore Power Connectors — 163
- Shore Power Pedestals Wiring — 166

Chapter 13 : AC Motors & Transformers — 169
- Single Phase Induction Motors — 170
- Dual Voltage Motors — 171
- Starting Capacitors — 172
- Three Phase Motors — 175
- Low and High Speed Motors — 178
- Three Phase Reverse Rotation — 179
- Contactors — 179
- Step Up and Down Transformers — 180
- Y and delta Connection — 182

Chapter 14 : Marine Generators — 183
- AC Generator Theory — 184
- Types of Excitation — 186
- AC Alternator — 186
- The frequency — 188
- Voltage Regulator — 190
- The Megger — 193
- How to Install a Marine Generator — 195
- Troubleshooting — 197
- The Generator Won't Crank — 197
- The Generator Start but Stop Suddenly — 198
- The Generator Crank but Doesn't Start — 199
- Low frequency — 200
- Under Voltage — 200
- Design the Harness — 201

Chapter 15 : Inverters and UPS's — 202
- Inverters — 203
- How Inverters Work — 204
- Types of Signal — 205
- Input/Output Power — 206
- Input Power Vs Output Power — 206
- Increasing the Output Capacity — 208
- UPS — 209
- Sizing the Battery Bank — 209
- Example of Inverter Batt. Bank Calculation — 210
- Fuses and Breakers AIC Rating — 211

Chapter 16 : Wiring Diagrams — 212
- Electrical & Electronic Components — 213
- Schematics — 214
- Wiring Diagrams — 214
- Rules — 215

Table of Contents

Chapter 17: Solar Power — **217**
The Solar Cell — 218
Photovoltaic Cell — 219
Solar Panel Configuration — 221
The Charger Controller — 222
Solar Panel System Calculation — 224
Boat Solar Panel System Configuration — 226

Chapter 18: Wind Power — **227**
Wind Mill Configuration — 228
AC / DC Wind Mills — 229
Types of Wind Mills — 229
How the Wind Mill works — 230

Chapter 19: Hybrid Vehicles — **231**
Hybrid and Electric Vehicles — 231
Hybrid Boats — 232
Electric & Hybrid — 233
Parallel Hybrid — 235

CHAPTER 1

The Nature of Matter

TOPICS

1.1 Atomic Number	11
1.2 Atomic Weight	11
1.3 Electronegativity	12
1.4 Ionization	13
1.5 Octet Ruler	13
1.6 Molecules	14
1.7 Oxidation and Corrosion	14
1.8 Static Electricity	15

Video Episode 1: The Nature of Matter

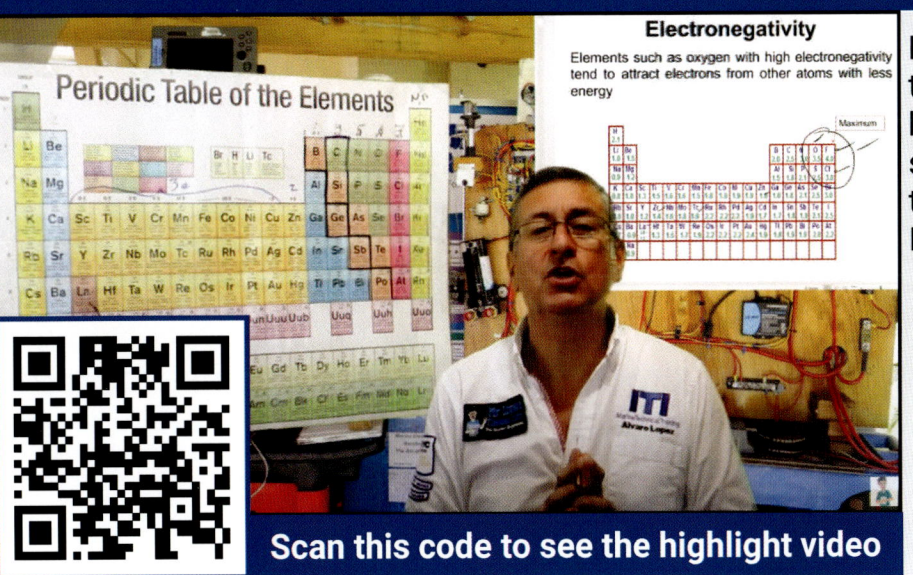

In this video you will learn the fundamentals of the electricity. how the electrons flow through some conductors and how the matter is organized In the periodic table.

Follow me

Scan this code to see the highlight video

Chapter 1: The Nature of Matter

The Nature of Matter

All matter is made of atoms composed of Protons, Neutrons, and Electrons. The center, or nucleus, of the atom is composed of positively charge protons and neutral neutrons. The outside of the atom has negatively charged electrons in various orbits.

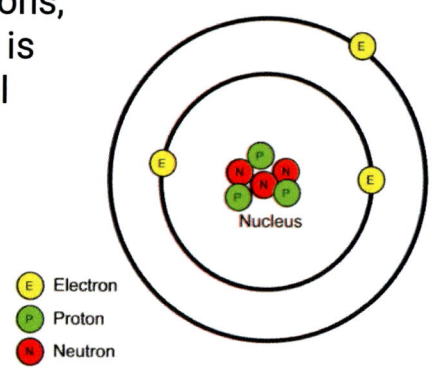

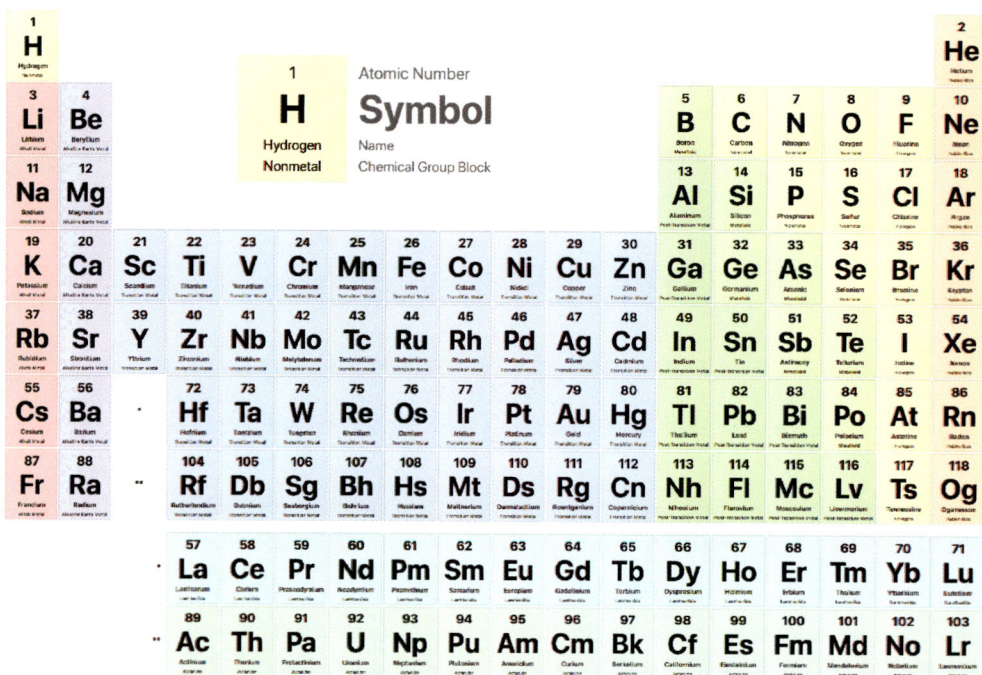

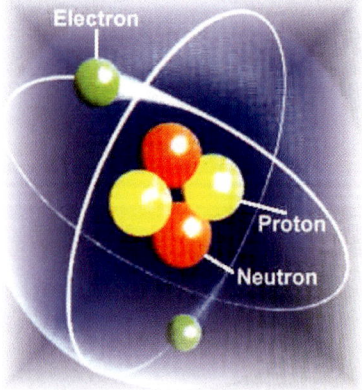

All atoms have the same number of protons (positively charged) and electrons (negatively charged). Therefore, all atoms have a neutral charge (the positive and negative charges cancel each other)

The Nature of Matter

Most atoms have approximately the same number of neutrons as they do protons or electrons, although this is not necessary, and the number of neutrons does not affect the identity of the element just his weight.

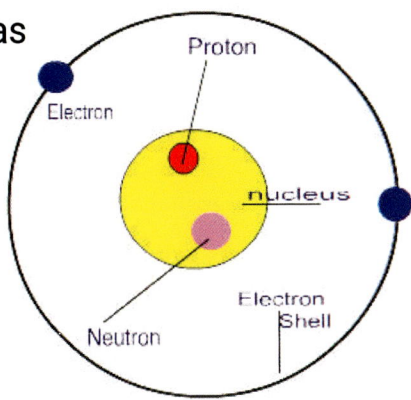

Atomic Number

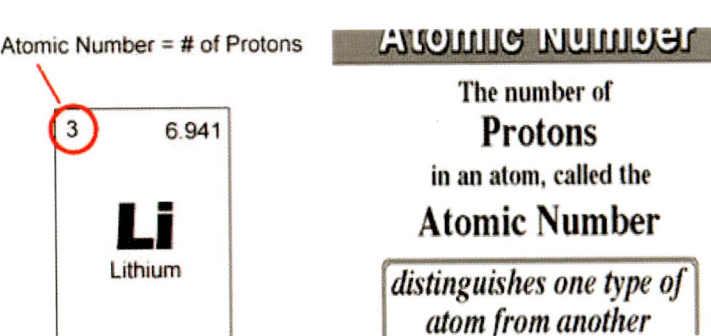

The Atomic Number of an element refers to the number of protons or electrons that make up an atom of that element.

Atomic Mass

The Atomic mass (Weight) of the atom is determined by the number of protons and neutrons in the nucleus (the electrons are so small as to be almost weightless)
The proton has 1836 times the mass of the electron, but exactly the same size charge, only positive rather than negative.

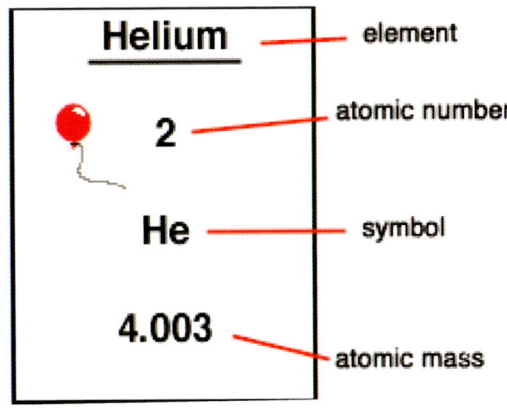

Electronegativity

Electronegativity is a measure of the tendency of an atom to attract a bonding pair of electrons.

Fluorine (the most electronegative element) is assigned a value of 4.0, and values range down to calcium and francium which are the least electronegative at 0.7

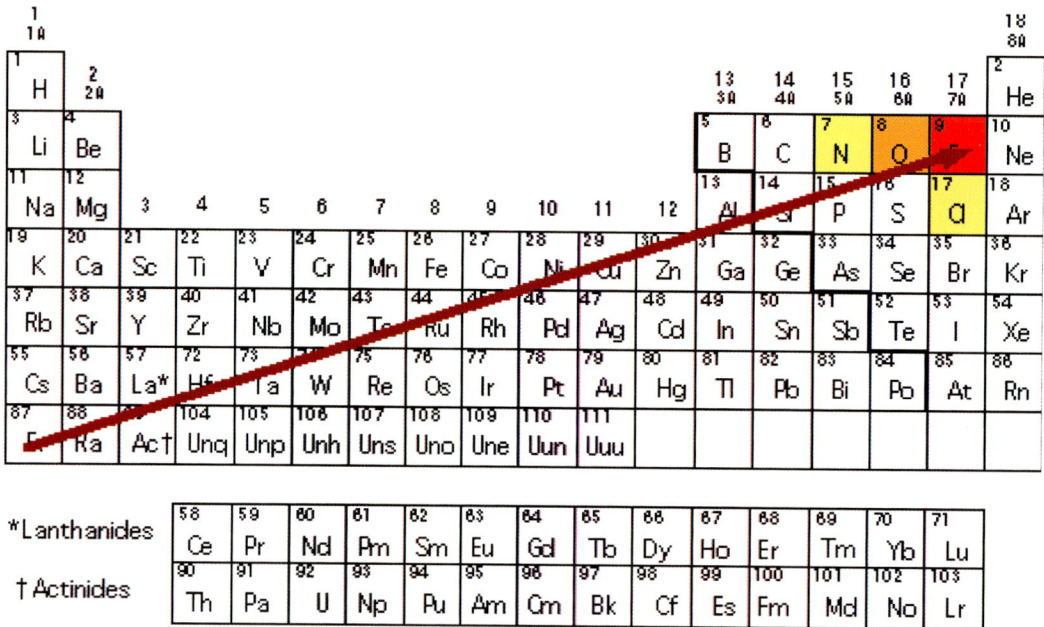

Elements such as oxygen with high electronegativity tend to attract electrons from other atoms with less energy.

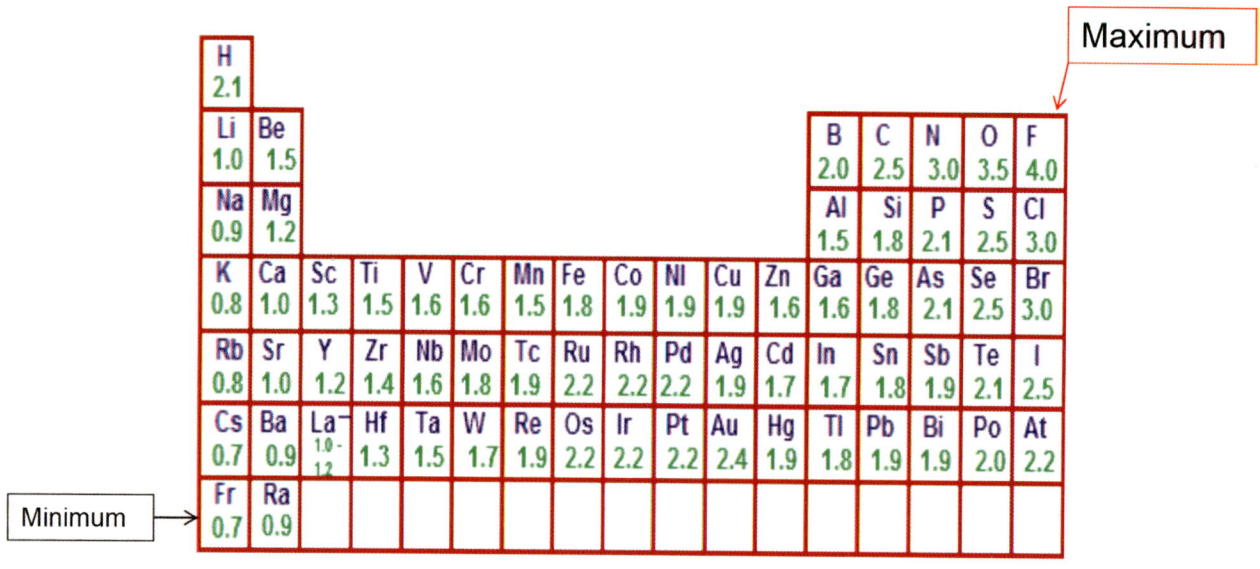

Ionization

When energy is added to an atom by some exterior force such as heat, friction or bombardment by other electrons, the atom becomes excited.

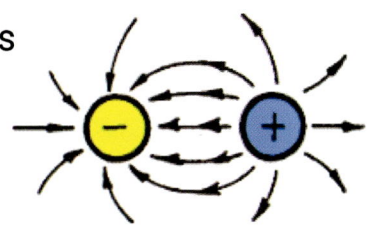

Ions are formed when atoms, or groups of atoms, lose or gain electrons.
Metals lose some of their electrons to form positively charged ions, e.g., Fe^{+2}, Al^{+3}, Cu^{+2}, etc.
Nonmetals located on the right side of the periodic table with high electronegativity try to gain electrons and form negatively charged ions, e.g.,
Cl^-, O^{-2}, S^{-2}, etc.

Hydrogen lose electrons, become positive

Water Molecule

Oxygen gain electrons, become negative

What is the Octet Rule?

The octet rule is a rule that states that all atoms "want to have" 8 or 0 outer electrons called valence electrons.
In this example, Oxygen with more electronegativity attracts two electrons from hydrogen atoms to complete 8 electrons in the outer orbit.
covalent bond In general, nonmetals located on the right side with high electronegativity try to gain electrons completing 8 electrons in the outer orbit. and metals located in the center and left side with less electronegativity tends to release electrons to get cero electrons in the outer orbit.

104.5°

The Water Molecule

The water molecule is a great example of a stable molecule due than the Hydrogens with lower energy than the oxygen release their electrons to get zero electrons in their outer orbit, then Oxygen with higher energy attracts those electrons getting eight electrons in his other orbit. This stable molecule does not react when external elements or substances are mixed.

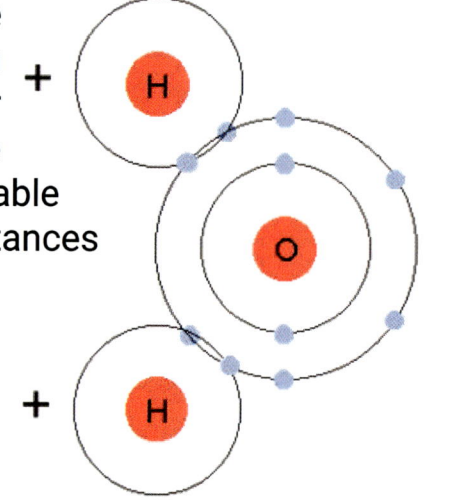

Oxygen molecule

This molecule results from the covalent bonding of two oxygen atoms which share each two atoms of its outer orbit. In this way the molecule is stable because both atoms have 8 electrons in the outer orbit.
This union is temporary for this reason oxygen layer is easily altered.

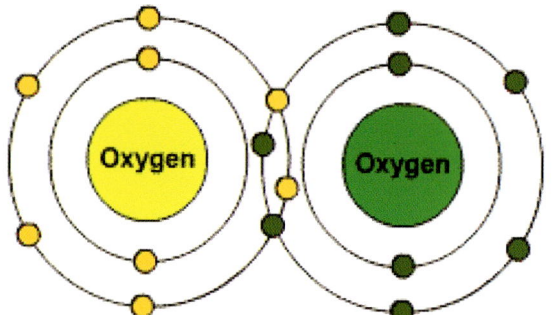

Oxidation

Is the interaction of oxygen molecules with other metals that are less electronegativity However, the oxygen attacks materials other than metals but that will be subject of study in our book dedicated to marine corrosion.

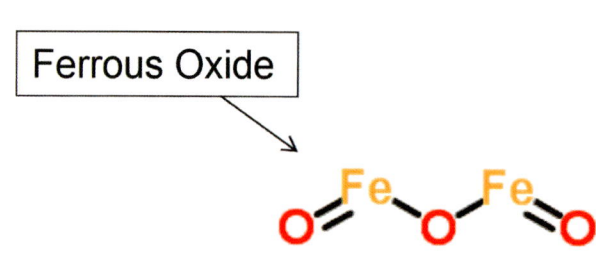

Ferrous Oxide

Al_2O_3

Prevention of Corrosion

Oxygen prefers to attack metals with only two electrons in its outer shell. In this way, the oxygen will have eight electrons and will be stable.

Zinc, with two electrons in its last orbit a good example, for this reason, it is used to avoid corrosion of other metals, due than the oxygen with six electrons prefer to attack Zinc instead of other metals.

Static Electricity

Static electricity is the result of an imbalance between negative and positive charges in an object.

These charges can build up on the surface of an object until they find a way to be released or discharged. One way to discharge them is through a circuit or through your body.

In simple terms static electricity is the accumulation of electrons coming from the windings of electric motors AC or DC. That static electricity could be store for long periods of time. In that case of the phenomenon of corrosion will be incremented in the boundary in between dissimilar metals. Once again this is a topic of our book about Marine Corrosion.

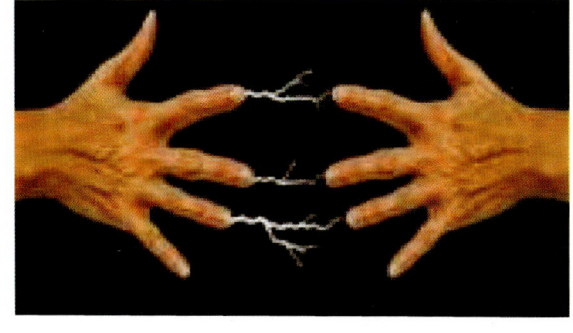

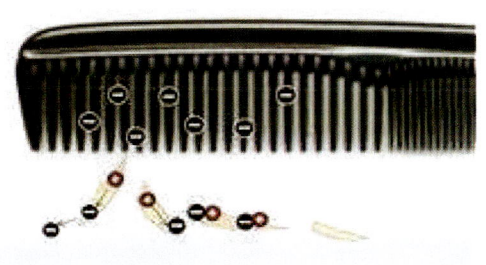

The rubbing of certain materials against one another can transfer negative charges, or electrons. For example, if you rub your shoe on the carpet, your body collects extra electrons. The electrons cling to your body until they can be released.

On dry winter days, static electricity can build up in our bodies and cause a spark to jump from our bodies to pieces of metal or other people's bodies. We can see, feel and hear the sound of the spark when it jumps.

Static Electricity

Some atoms hold on to their electrons more tightly than others do (More or less electro-negativity)

The only way to drain those static currents is through the grounding conductor (green color). Each metallic element in a boat should be connected to the grounding conductor (Bonding) to drain those currents and prevent corrosion.

The static charges remain on an object until they either bleed off to ground or are quickly neutralized by a discharge.

Static electricity can be contrasted with dynamic current, which can be delivered through wires as a power source. Although charge exchange can happen whenever any two surfaces come into contact and separate, a static charge only remains when at least one of the surfaces has a high resistance to electrical flow (an electrical insulator).

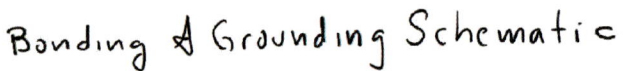

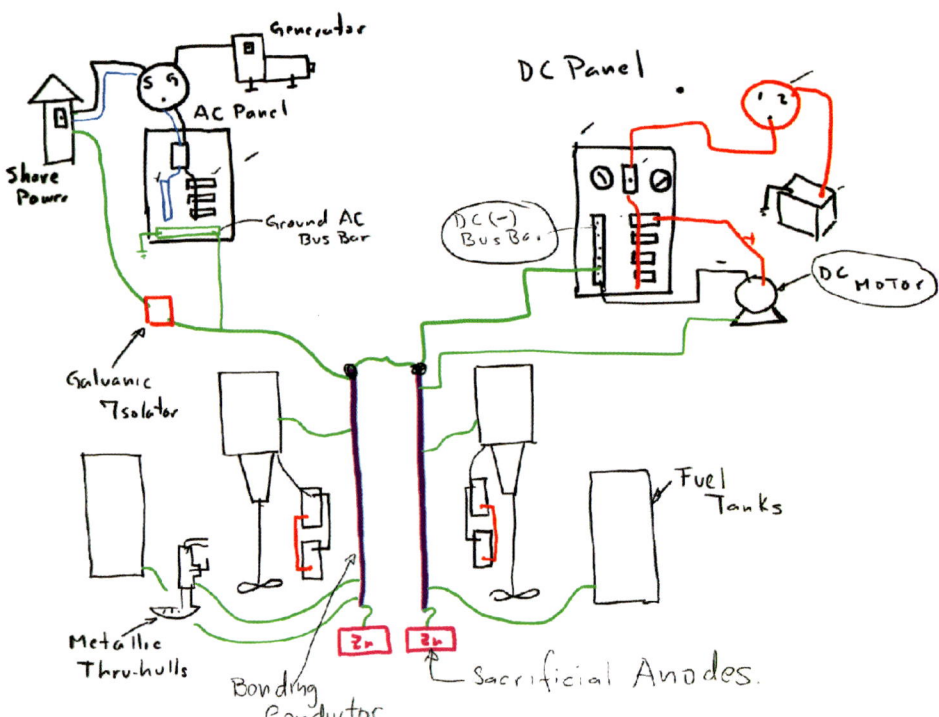

CHAPTER 2

The Electrical Current

TOPICS

2.1 The flow of charges	18
2.2 The Ohm Law	19
2.3 The Potential Differential	20
2.4 Types of Current	20
2.5 DC Current	21

Video Episode 1: The Electrical Current

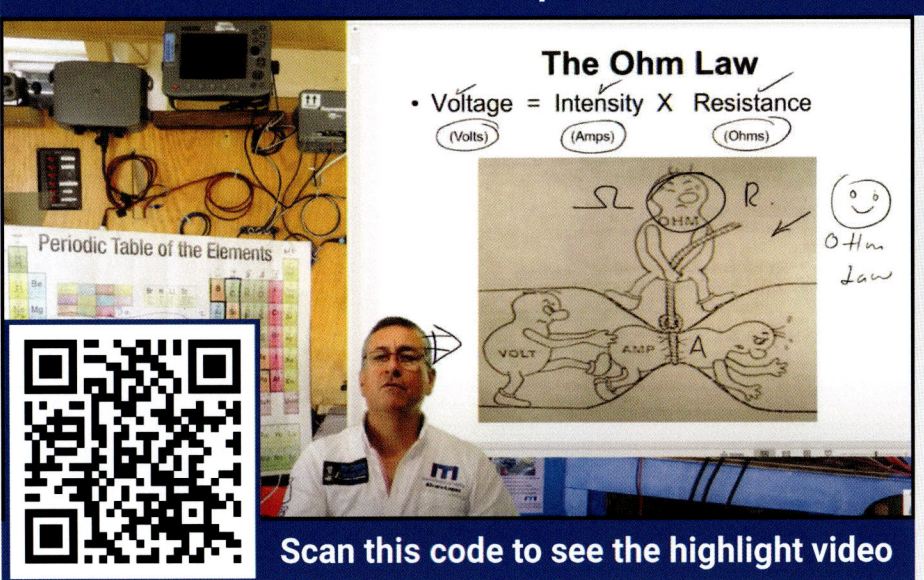

Scan this code to see the highlight video

In this Chapter we are going to study the different types of current used in a boat wiring. Using the Ohm law we will calculate parameters such as: Intensity (Amp), Resistance (Ohm), Voltage (Volt) and Power (Watts).

Follow me

The Current

The flow of charge is called the current and it is the rate at which electric charges pass though a conductor.
The charged particle can be either positive or negative.
In order for a charge to flow, it needs a push (a force) and it is supplied by voltage, or potential difference. The charge flows from high potential energy to low potential energy
The Voltage is measured in Volts =V

The flow of Charges

In a battery the flow of electron starts from the positive terminal and finish in the negative post, however, the flow of electricity occurs in the opposite way. Each time an electron migrates from the positive post then a hole is created and this terminal will be more and more positive because more and more electrons migrate, this is why at the end the positive terminal will be eroded. That is the reason why the positive terminal is also called "Sacrificial Anode" and of course, the positive terminal is always called the Anode, consequently, the negative terminal will be called the Cathode.

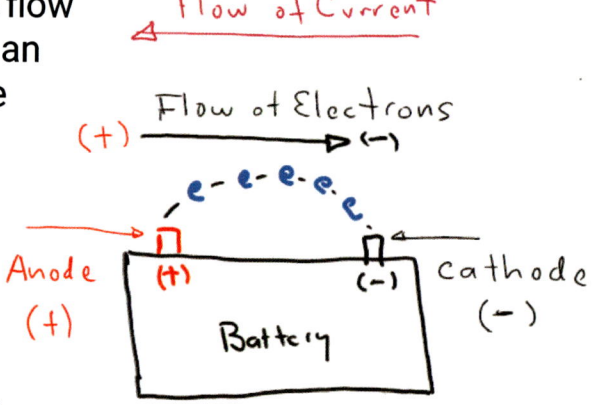

Flow of Current Vs Electron Movement

Until now nobody has been able to explain this phenomenon, most of the books conclude that Benjamin Franklin made a mistake when naming positive and negative charges.
In my opinion, the answer is in <u>the third law of Newton</u> which states that: For every action, there is an equal and opposite reaction. This is why the flow of current travels in opposite direction than the flow of electrons.
When you sit in your chair, your body exerts a downward force on the chair and the chair exerts an upward force on your body.

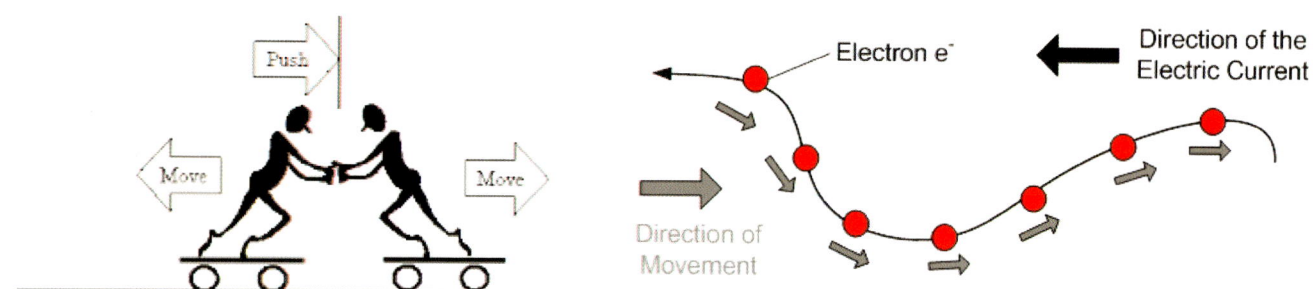

The Current

The unit for current is Ampere [Amp per Hour]. The rate of flow of electric charge is called Electric Current and is measured in Amperes per hour.
To measure the flow of charges the circuit should be activated and the measurement instrument should be placed in series.

Current Formula
I (Amp) = Volt / R (Ohms)

According to Ohm's Law: the Voltage **V** in an ideal circuit is proportional to the applied Intensity **I**, times the Resistance **R**

$$V = \downarrow I \times R \uparrow$$

If the voltage remains constant then when the resistance increases, the current decreases.

In other words, for a fixed Resistance (R), the greater the Voltage (V) across a Resistor, the more the Current (I) flowing through it; for a fixed Voltage across a Resistor, the more the Resistance of the Resistor, the less the Current flowing through it.

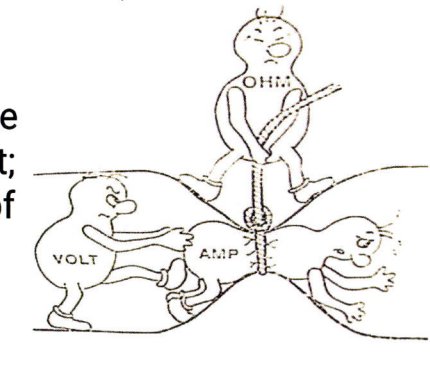

$$\uparrow I = \uparrow V/R$$

The Ohm Law

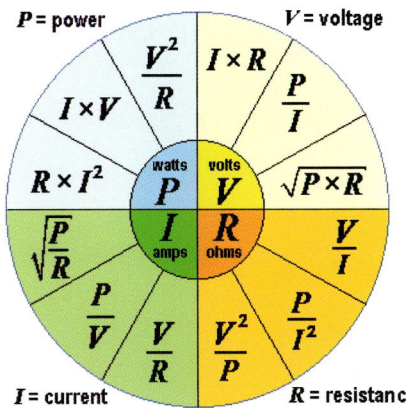

Voltage (Volts) = Intensity (Amps) X Resistance (Ohms)

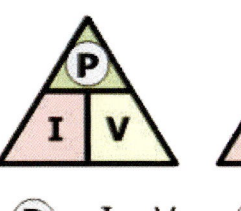

$(P) = I \times V \quad (I) = \dfrac{P}{V} \quad (V) = \dfrac{P}{I}$

Electric Power

Electric power is the rate of energy consumption in an electrical circuit.
The electric power is measured in units of **watts**

$$P = V \times I = \frac{V^2}{R} = I^2 R$$

1 KW = 1000 Watts
1 Watt = 1 Volt x 1 Amp-Hr

The Potential Difference

Suppose **A** has a potential of 12 V and **B** has a potential of 2 V. There is a potential difference. A has higher potential energy than B, and it means there is voltage.
The potential difference is VA - VB = 12 - 2 = 10 V.

Suppose **A** and **B** has a potential of 10 V. There is no potential difference between two plates.
The potential difference is VA - VB = 10 - 10 = 0 V Therefore, it has no voltage, and it means no flow of charge.

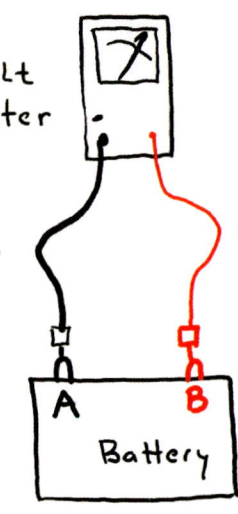

Types of Current

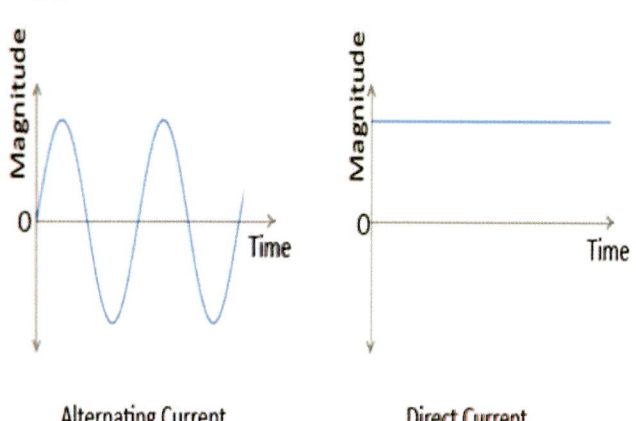

There are two types of electric current: direct current (DC) and alternating current (AC).
The electrons in direct current flow in one direction. The current produced by a battery is direct current.
The electrons in alternating current flow in one direction, then in the opposite direction—over and over again.

DC Current

Direct current (DC) is the unidirectional flow of electrons.
Direct current is produced by sources such as Batteries, Thermocouples, Solar Cells and Alternators.

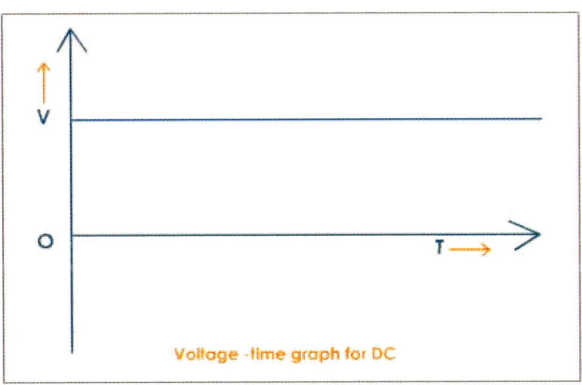
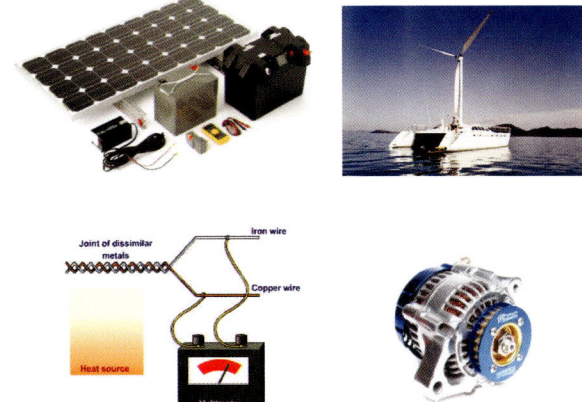

AC Current

If a machine is constructed to rotate a magnetic field around a set of stationary wire coils with the turning of a shaft, AC voltage will be produced across the wire coils as that shaft is rotated. In a Typical recreational Boat: There are three sources of AC current: Generators, Inverters and Shore Power.

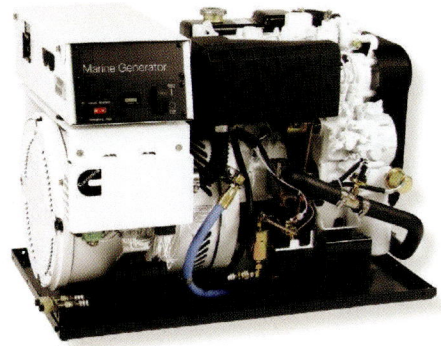

DC Current from AC

Direct current may be obtained from an alternating current supply by use of a current-switching arrangement called a **rectifier.**
A battery charger is a sophisticated rectifier in chapter 6 we will talk deeply about those types of units.

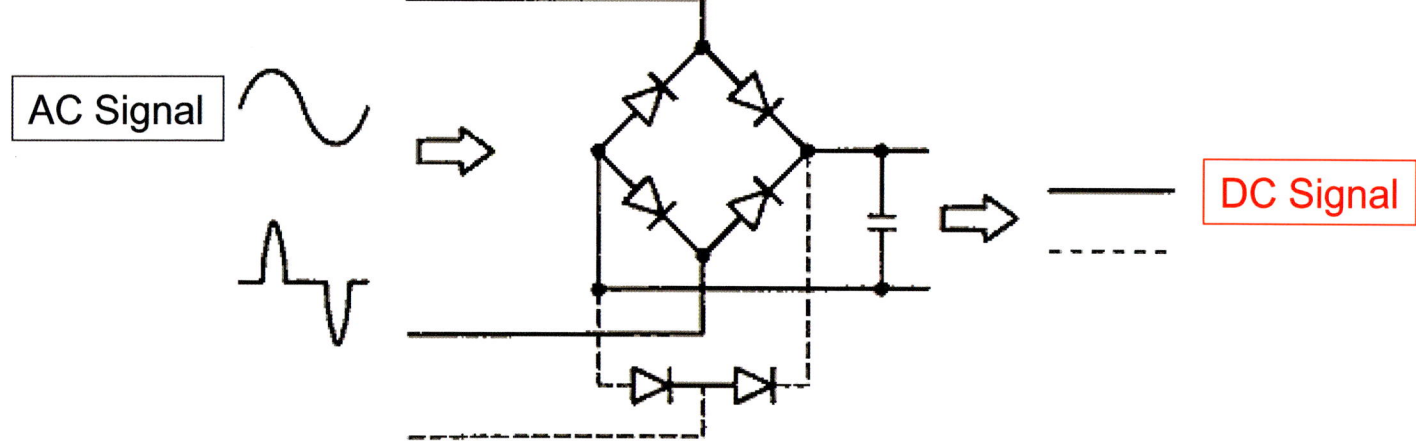

AC Current from DC

Direct current may be made into alternating current with an **Inverter.** This is a complex process. We will study the Inverters (Chapter 15) and how they are more accurate every day, due to the solar designs.

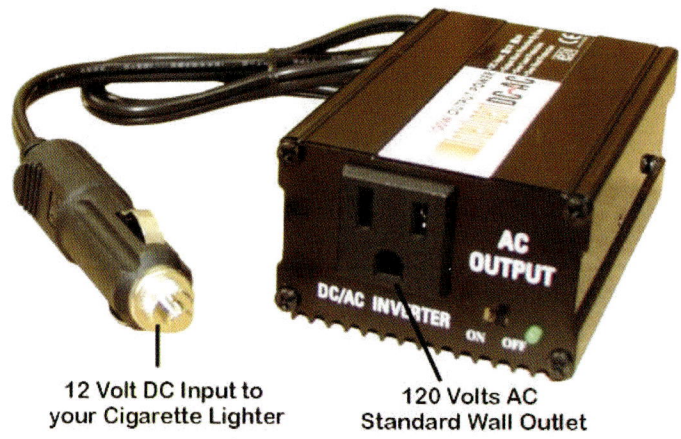

CHAPTER 3

Marine Batteries

TOPICS

3.0 What is a Battery	24	3.10 Nickel-Cadmium Batteries	33
3.1 Marine Battery Classifications	25	3.11 Amperes – Hour	33
3.2 Deep Cycle Batteries	26	3.12 State of Charge	33
3.3 Cranking Batteries	27	3.11 Cold Cranking Amps	34
3.4 Bank of Batteries	28	3.13 Battery load Tester	34
3.5 The Electrolyte	29	3.14 Storage Capacity	36
3.6 Charging Process	29	3.15 Battery Terminals	37
3.7 Specific Gravity	30	3.16 Battery Tips	38
3.8 The hydrometer	31		
3.9 Lithium-Ion Batteries	32		

Video Episode 1: Marine Batteries configuration and Diagnosis I

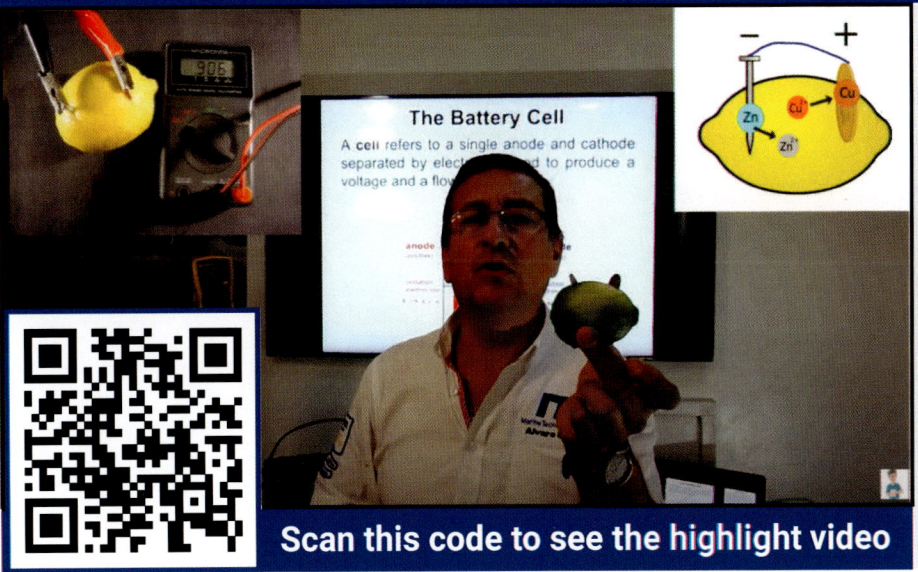

In this video you will learn the fundamentals about Marine Batteries. In the video you will find a lot of recommendation about the service on the batteries, the procedure to check the battery condition and the considerations in order to configure a group of batteries.

Scan this code to see the highlight video

Follow me

Chapter 3: Marine Batteries

What is a Battery?

Is a group of cells connected in series to create a flow of electrons through an electrolyte.

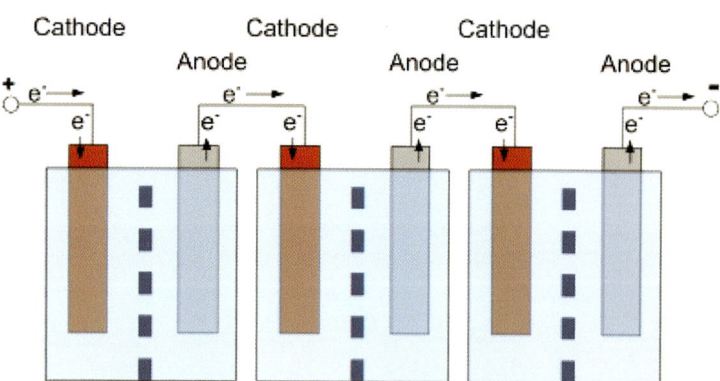

The Battery Cell

A cell refers to a single anode and cathode separated by an electrolyte (Dry, Wet, and Gel) used to produce a voltage and a flow of charges.
The configuration of modern battery banks using individual cells is more polar every day. The individual Lithium Ion cells can be connected in series-parallel arrays to get a higher Voltage and Intensity.

Battery Bank for Hybrid Car vs Battery Bank for Fully Electric Car

The Anode

The anode is always the positive terminal in a battery. The anode becomes in a positive terminal because the electrons are migrating through the electrolyte, each time than an electron goes out, the anode is more positive because one negative charge (electron) is going out.

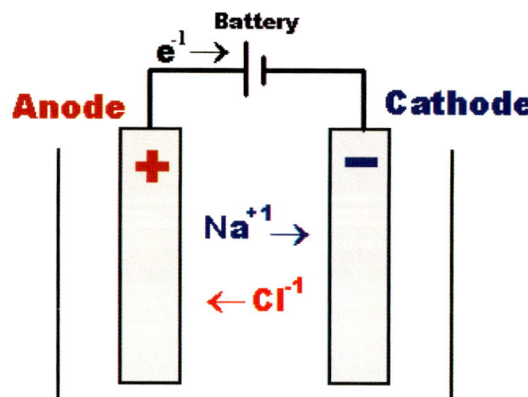

A Sacrificial Anode

Each time than an electron migrate from the anode, a hole is produced on the structure of the anode.
For that reason, the anode is commonly called a sacrificial metal. At the end of the day the anode will be fully eroded and consequently corroded.

Marine Batteries Classification

According to the purpose batteries are classified in:

- Starter
- Deep Cycle
- Dual - Purpose

According to the Type are classified in

- Wet / Flooded   (Lead Acid)
- Gel
- Absorbed Glass Material (AGM)
- Lithium- Ion / Lithium Phosphate

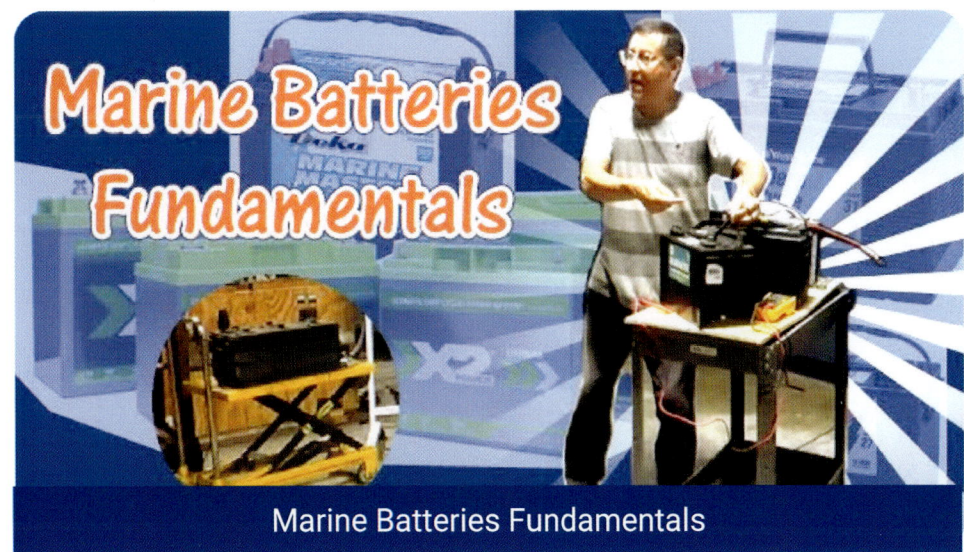

Marine Batteries Fundamentals

Deep Cycle Batteries

The thicker the lead plates, the longer the life span, all things being equal. Battery weight is a simple indicator for the thickness of the lead plates used in a battery.
The heavier a battery for a given group size, the thicker the plates, and the better the battery will tolerate deep discharges
Some "Marine" batteries are sold as dual-purpose batteries for starter and deep cycle applications.

Deep Cycle Batteries are also common in golf carts and large solar power systems (the sun produces power during the day and the batteries store some of the power for use at night).

Crank or Deep-Cycle ?

A Cranking battery is designed to provide a very large amount of current (Amp) for a short period of time
·   Once the engine starts, the alternator provides all the power that the engine needs, so a cranking battery may go through its entire life without ever being drained more than 20 percent of its total capacity.
A deep cycle battery is designed to provide a steady amount of current over a long period of time
·   A deep cycle battery is also designed to be deeply discharged over and over again.

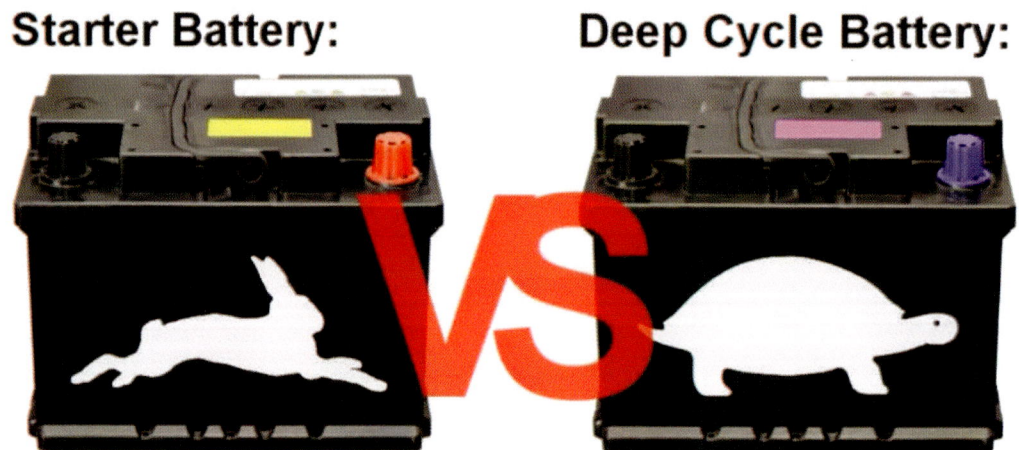

Starter / Cranking Batteries

This type of Batteries have a thinner and closer plates internally, providing more surface area to give that higher,
one-time discharge power during the cranking process.
They have many thin lead plates which allow them to discharge a 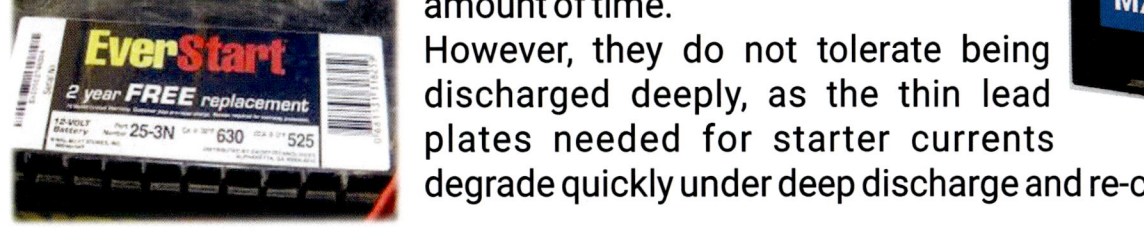 lot of energy very quickly for a short amount of time.
However, they do not tolerate being discharged deeply, as the thin lead plates needed for starter currents
degrade quickly under deep discharge and re-charging cycles.

Most starter batteries will only tolerate being completely discharged a few times before being irreversibly damaged.

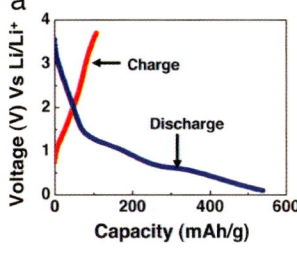 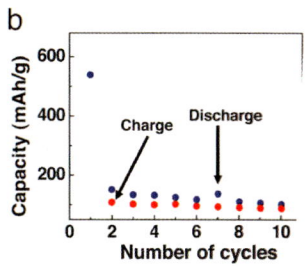

Deep Cycle Batteries

Deep Cycle batteries have thicker lead plates that make them tolerate deep discharges better.
They cannot dispense charge as quickly as a starter battery but can also be used to start small combustion engines. You would simply need a bigger deep-cycle battery than if you had used a dedicated starter-type battery instead.
Deep cycle batteries are commonly used in House Battery Banks, where you need storage 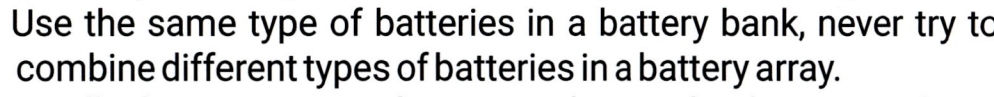 and delivery of the highest amount of Amp at low discharge rates.

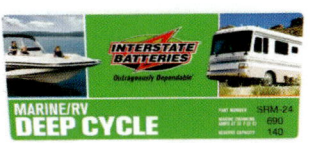
Use the same type of batteries in a battery bank, never try to combine different types of batteries in a battery array.
Finally does not matter how many battery banks are in a boat; the negative terminal of all of those banks should converge in a common DC negative bus bar; then the negative bus bar should be connected to the main Bonding conductor. This is the only way to assure that the signals travel from different places in a boat.

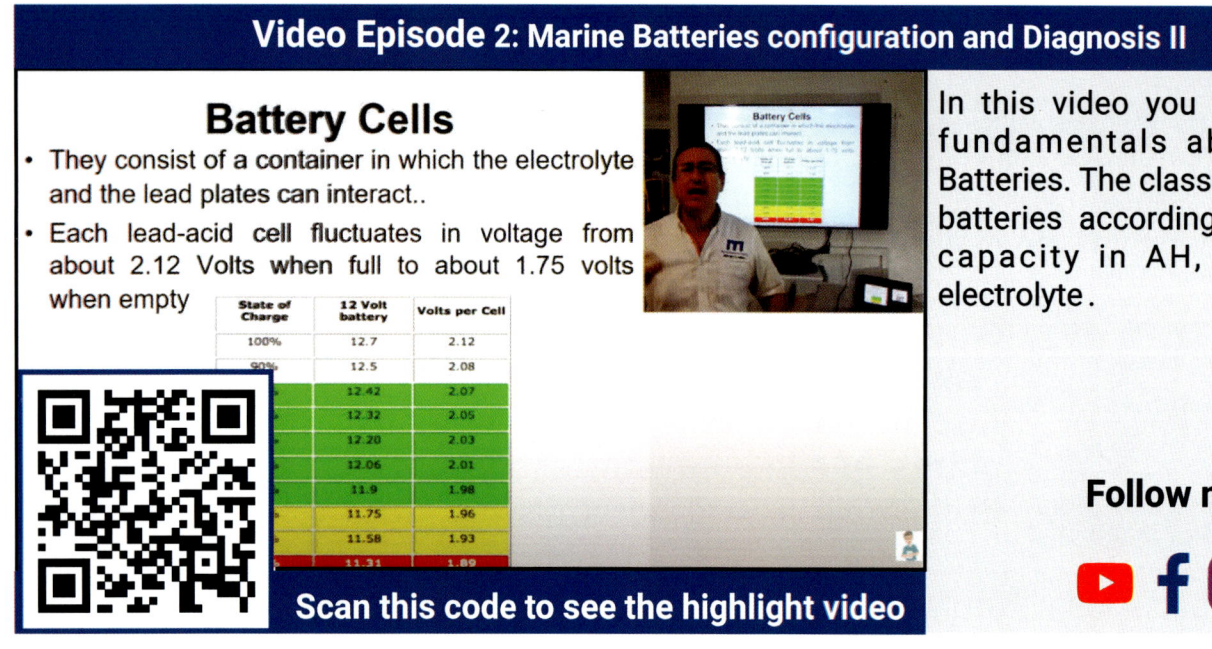

Bank of Batteries

Normally in a medium pleasure yacht there are minimum five groups of batteries

- Cranking Batteries
- House Batteries
- Generator Batteries
- Bow & Stern Thruster
- Electronics

The negative terminal from each group shall be connected together into a negative bus bar and the bar with the bonding system.
It is recommended that in each bank all batteries have the same characteristics such as

- Voltage (12 or 24V)
- Type of Material
- Capacity (Amp/Hour)

Keep in mind the recommendations given in this book about:

- Battery Chargers
- Battery Isolators
- Alternators

Flooded / Lead Acid Batteries

A lead-acid battery is an electrical storage device that uses a reversible chemical reaction to store energy.
It uses a combination of lead plates or grids and an electrolite consisting of a diluted sulfuric acid to convert electrical energy into potential chemical energy and back again.

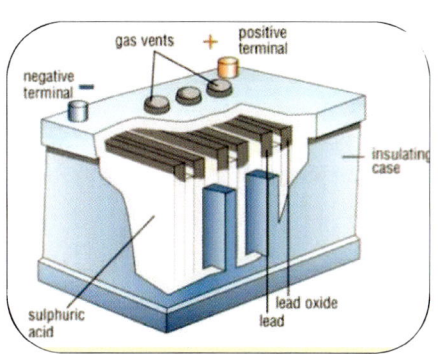

The Electrolyte

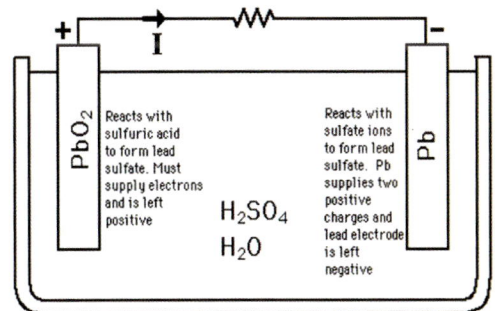

The electrolyte of lead-acid batteries is hazardous to your health and may produce burns and other permanent damage.

Battery electrolyte is a mixture of water (density 1.000 grams/cubic centimeter) and Sulfuric Acid (density 1830 grams/cubic centimeter).
Since the flow of electricity into and out of a battery results in either generation of sulfuric acid or loss of sulfuric acid
The level of the charge in the battery is a direct, linear function of the density of the electrolyte.

Charging Process

The conversion of energy from chemical to electrical is known as the charging. And when the electric power changes into chemical energy then it is known as discharging of the battery.

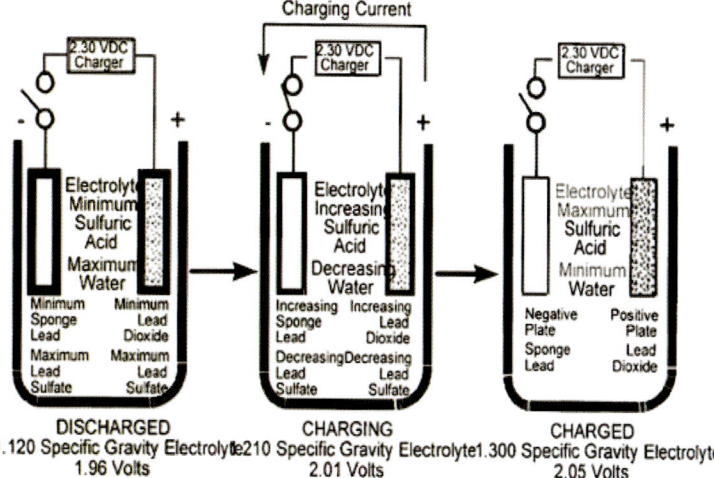

Charging Process

The lead-acid battery mainly uses two types of charging methods namely the constant voltage charging and constant current charging.

The constant voltage process is the most common method of charging the lead acid battery. It reduces the charging time and increases the capacity up to 20%. But this method reduces the efficiency by approximately 10%.

Battery Cells

They consist of a container in which the electrolyte and the lead plates can interact.

Each lead-acid cell fluctuates in voltage from about 2.12 Volts when full to about 1.75 volts when empty.

State of Charge	12 Volt battery	Volts per Cell
100%	12.7	2.12
90%	12.5	2.08
80%	12.42	2.07
70%	12.32	2.05
60%	12.20	2.03
50%	12.06	2.01
40%	11.9	1.98
30%	11.75	1.96
20%	11.58	1.93
10%	11.31	1.89

Specific Gravity

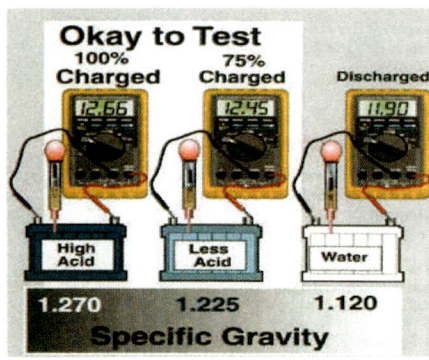

The electrolyte solution consists of 65% water and 35% sulfuric acid.

The specific gravity or weight of this solution increases as the battery charges and decreases as the battery discharges.

As the battery discharges the sulfur moves away from the solution and toward the plates.

The opposite is true as the battery is charged, the sulfur returns to the electrolyte solution.

The specific gravity of the electrolyte depends on this 65% to 35% ratio for the chemical reaction necessary create the electrons in the battery.

As the temperature drops the electrolyte contracts increasing the specific gravity and changing the reading.

State of Charge	12 Volt	6 Volt	Specific Gravity
100%	12.9	6.4	1.265
75%	12.4	6.2	1.225
50%	11.9	6.0	1.190
25%	11.4	5.8	1.155
Discharged	10.5	5.5	1.120

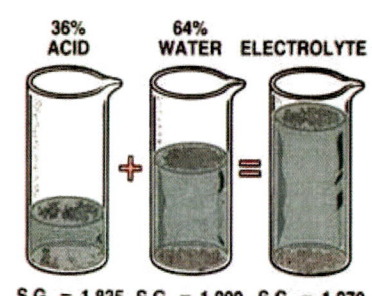

S.G. = 1.835   S.G. = 1.000   S.G. = 1.270

The Hydrometer

Use a hydrometer to measure the specific gravity of the electrolyte solution in each cell. It's a tool used to measure the density or weight of a liquid compared to the density of an equal amount of water.
A lead acid battery cell is fully charged with a specific gravity of 1.265 at 80° F

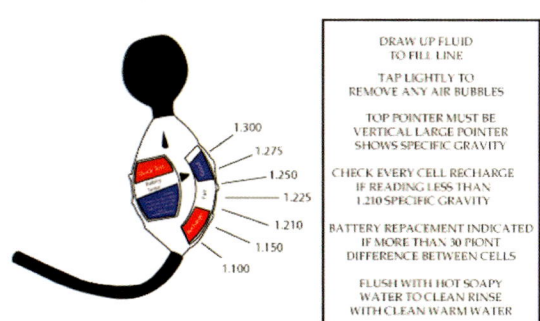

CELL READINGS	PERCENT CHARGED
1.270	100 %
1.230	75%
1.190	50%
1.145	25%
1.100	0%

Valve regulated Lead Acid Batteries

The two most common VRLA batteries used today are the *Gel* and *Absorbed Glass Mat* (AGM) variety.
Gel batteries feature an electrolyte that has been immobilized using a gelling agent like fumed silica.
AGM batteries feature a thin fiberglass felt that holds the electrolyte in place like a sponge. Neither AGM or Gel cells will leak if inverted, pierced, etc. and will continue to operate even under water.

Lithium – Ion Batteries

Lithium batteries have a high energy density and are excellent for deep cycle applications. Compared to flooded batteries, lithium batteries deliver a savings of up to 70 percent in volume and weight, with three times as many charging cycles.

High Quality Lithium Ion Phosphate Batteries

Lithium batteries handle gigantic amounts of current, so they can be recharged faster than any other type.
Lithium is in use onboard high-performance offshore racing sailboats and others whose owners demand extreme weight savings and bleeding-edge performance and are willing to pay premium prices.

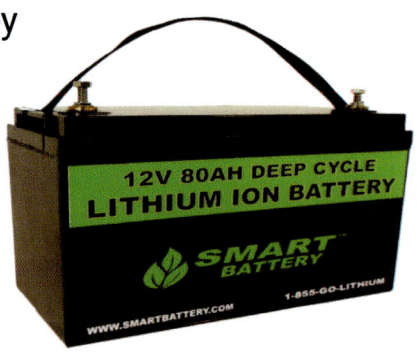

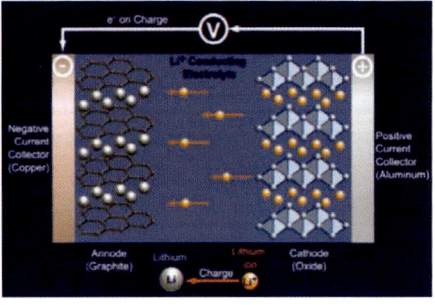

Lithium batteries offer savings of up to 70% in volume and weight compared to traditional lead-acid batteries, with three times as many charging cycles (2000 full cycles).
They are equipped with a Battery Management System (BMS), which automatically compensates for any imbalance between the cells.

Nickel-Cadmium Batteries

The nickel cadmium batteries are the most expensive batteries used in marine applications due to the high cost of the metals involved and the alkaline electrolyte. They have a life around 15 years with a minimum maintenance.

Amperes / Hour

A battery with a capacity of 1 amp-hour should be able to continuously supply a current of 1 amp to a load for exactly 1 hour
For example, an average Marine battery might have a capacity of about 70 amp-hours, specified at a current of 3.5 amps. This means that the amount of time this battery could continuously supply a current of 3.5 amps to a load would be 20 hours (70 amp-hours / 3.5 amps.

Reserve Capacity (RC)

Reserve Capacity (RC) is the number of minutes a fully charged battery at 80 o F (26.7 o C) is discharged at 25 amps before the voltage falls below 10.5 volts.
To convert Reserve Capacity (RC) to Ampere-Hours at the 25-amp rate, multiple RC by 0.4167. More ampere-hours (or RC) are better in every case
RC and Ah are NOT one in the same as I often see people use them interchangeably.

State of Charge

The State of Charge describes how full a battery is. The exact voltage to battery charge correlation is dependent on the temperature of the battery.
Cold batteries will show a lower voltage when full than hot batteries. This is one of the reasons why quality alternator regulators or high-powered charging systems use temperature probes on batteries.

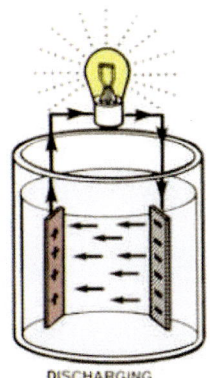
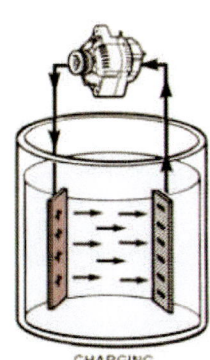

Depth of Discharge (DOD)

The Depth of Discharge (DOD) is a measure of how deeply a battery is discharged.
When a battery is 100% full, then the DOD is 0%. Conversely, when a battery is 100% empty, the DOD is 100%.
Most battery manufacturers advocate not discharging their batteries more than 50% before re-charging them.

Cold Cranking Amps (CCA)

CCA is critical for good cranking ability. It's the number of amps a battery can support for 30 seconds at a temperature of 0 degrees F until the battery voltage drops to unusable levels.
A 12V battery with a rating of 600 CCA means the battery will provide 600 amps for 30 seconds at 0 degrees before the voltage falls to 7.20 V (six cells).

Battery Load Tester

If you do not use a load specific device when you load your battery, you may experience false positive readings on weak batteries, and this could lead to many problems down the road.

Battery Load Tester

A proven battery tester that applies a precise load and compensates for temperature variance will make all the difference in the world. Many of the battery load devices without this feature will pass a weak battery, because they are not exercising the batteries potential during the test.

Battery Load Tester

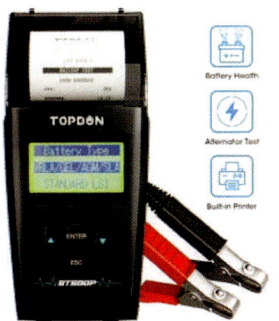

Many other testers will only assess the voltage and not even apply a predetermined load. If there is any voltage whatsoever the tester will give a false positive and false sense of security to say the least.

Depending on your selection you may have a panel range to perform adjustments of 900 CCA to 1200 CA and up to 125 degrees in temperature to adjust the temperature controls. As an alternative the settings may be in AHR or Ampere Hours.

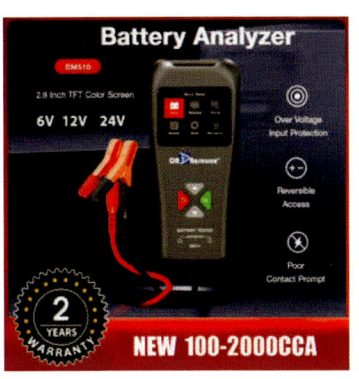
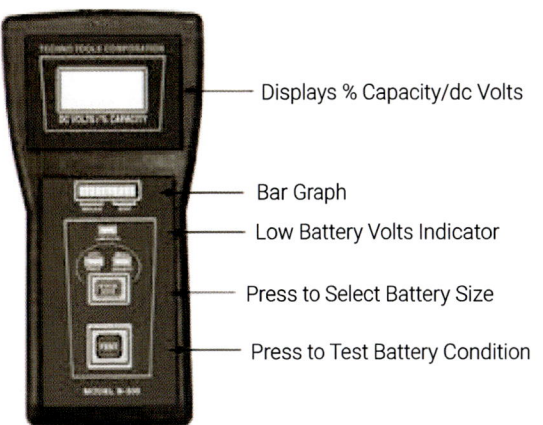

- Displays % Capacity/dc Volts
- Bar Graph
- Low Battery Volts Indicator
- Press to Select Battery Size
- Press to Test Battery Condition

Video Episode 3: Marine Batteries Series & Parallel configuration

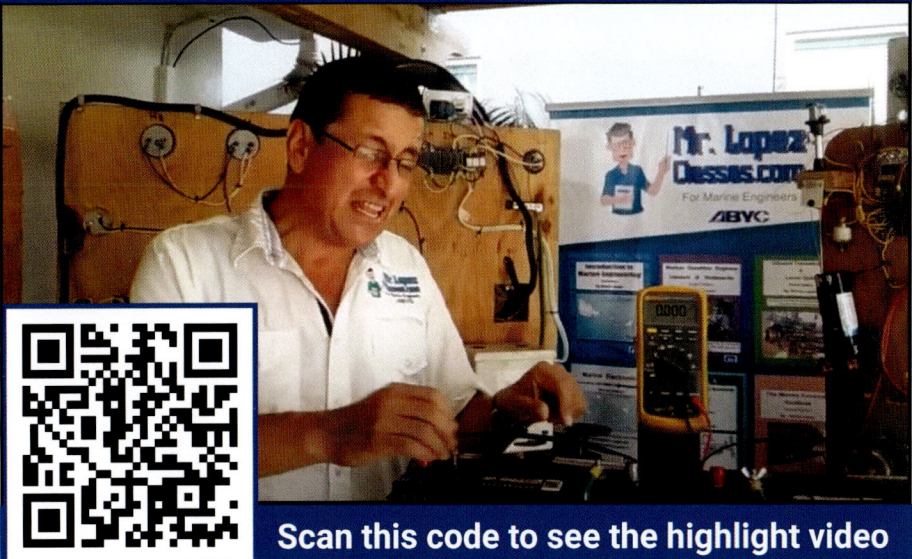

Scan this code to see the highlight video

In this video you will learn the procedure to configure arrays of batteries in series, parallel, combined and a lot of recommendations in order to upgrade battery banks or replace batteries in an old group of batteries.

Follow me

Battery Storage Capacity

The **Amp-hour (Ah)** Capacity of a battery tries to quantify the amount of usable energy it can store at a nominal voltage.

Storage capacity is additive when batteries are wired in parallel but not if they are wired in series.

When two 6V, 100Ah batteries are wired in Series, the voltage is doubled but the amp-hour capacity remains 100Ah (Total Power = 1200 Watt-hours).

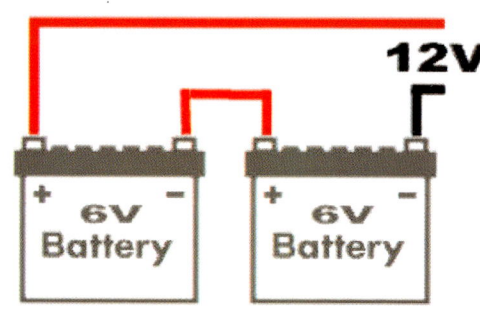

You may decide to wire batteries in series because a single 12V battery with the right storage capacity is simply too heavy, unwieldy, or awkward to lift into place. Batteries consisting of fewer cells (and hence lower voltage) in series can provide the same storage capacity yet be portable.

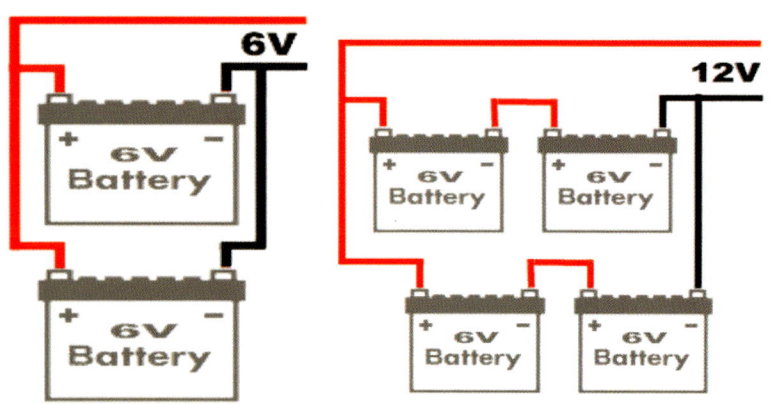

Two 6V, 100Ah batteries wired in Parallel will have a total storage capacity of 200Ah at 6V (or 1200 Watt-hours).

Battery banks wired in Series-Parallel are even more complicated. Here, four 6V cells are wired in two "strings" of 12VDC that were then wired in parallel. Using 6V, 100Ah batteries, this system will have a storage capacity of 200Ah at 12V or 2,400Wh.

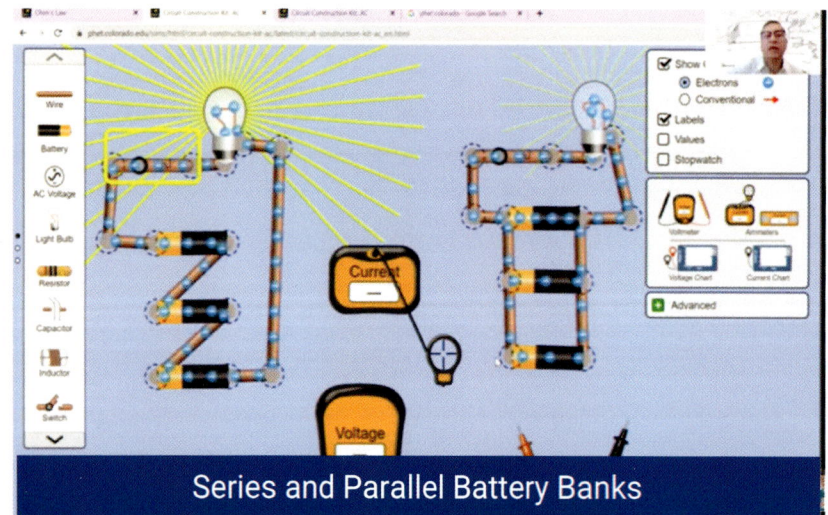

Series and Parallel Battery Banks

Temperature Effects on Batteries

Battery capacity (how many amp-hours it can hold) is reduced as temperature goes down, and increased as temperature goes up.
This is why your boat battery dies on a cold winter morning, even though it worked fine the previous afternoon.
If your batteries spend part of the year shivering in the cold, the reduced capacity has to be taken into account when sizing the system batteries. The standard rating for batteries is at room temperature - 25 degrees C (about 77 F).

Battery Terminals

Marine batteries typically have two posts, a 3/8"-16 threaded post for the **Positive** terminal, and a 5/16"-18 threaded post for the **Negative** terminal.

Wing nuts are not ABYC approved

Battery Tips

Stay with one battery chemistry (flooded, gel or AGM) Each battery type requires specific charging voltages. Mixing battery types can result in under- or over-charging. This may mean replacing all batteries on board at the same time.
Never mix old batteries with new ones in the same bank. While it seems like this would increase your overall capacity, old batteries tend to pull down the new ones to their deteriorated level.

Battery Tips

Regulate charge voltages based on battery temperature and acceptance (manually or with sensing) to maximize battery life and reduce charge time. Ensure that your charging system is capable of delivering sufficient amperage to charge battery banks efficiently.
This generally means an alternator with 25% to 40% as many amperes as the capacity of your entire battery bank.

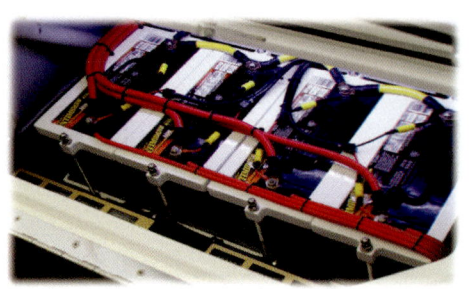

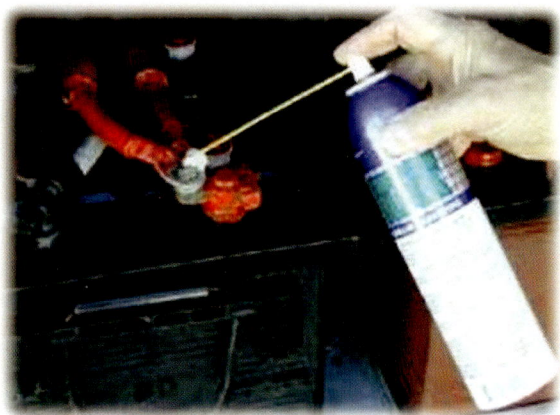

Keep batteries clean, cool and dry.
Check terminal connectors regularly to avoid loss of conductivity.
Add battery electrolyte to flooded lead acid batteries when needed. Keep them charged.
Clean corrosion with a paste of baking soda and water.

Tips to Connect Boat Battery Banks

CHAPTER 4

Conductors and Isolators

TOPICS

4.0 Isolators	40
4.1 Conductors	40
4.2 Wire Connectors	43
4.3 Electrical Conductors	44
4.4 Circular Mils	44
4.5 Voltage Drop	45
4.6 Sizing AWG wire using ABYC Tables	47
4.7 Marine Wire Gauge	52
4.8 Marine Wire Color Code	53
4.9 Wire Sizes	55
4.10 Marine Duplex/Triplex wire	55
4.11 Continuity	58

Video Episode 1: Conductors - Isolators - Terminals & Wire Gauge

Scan this code to see the highlight video

In this video you will learn the fundamentals about conductors and Isolators, then you will find the procedure to calculate the wire gauge according with ABYC standards and finally you will learn the process to install wire terminals using the appropriate tools.

Follow me

Conductors and Insulators

Metals such as copper typify conductors, while most non-metallic solids are said to be good insulators, having extremely high resistance to the flow of charge through them.

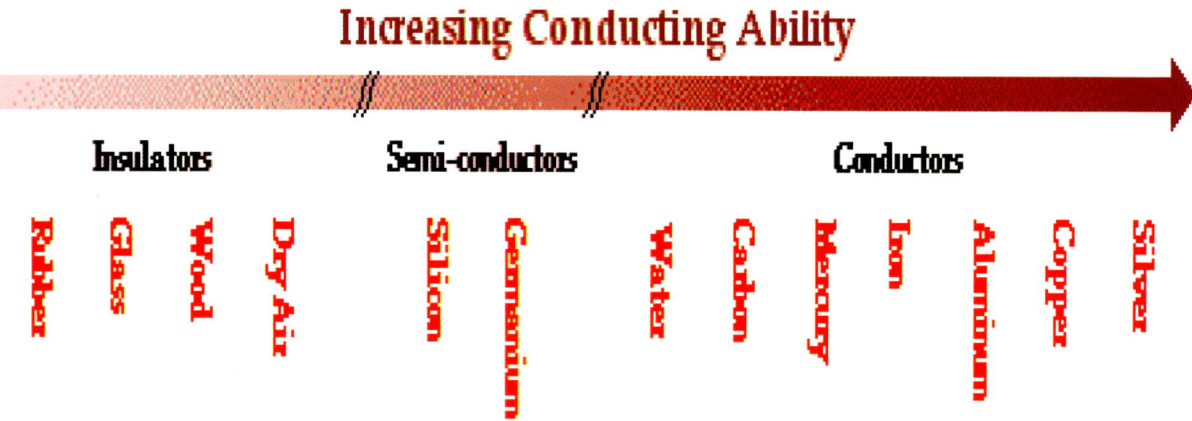

"**Conductor**" implies that the outer electrons of the atoms are loosely bound and free to move through the material. Most atoms hold on to their electrons tightly and are insulators.

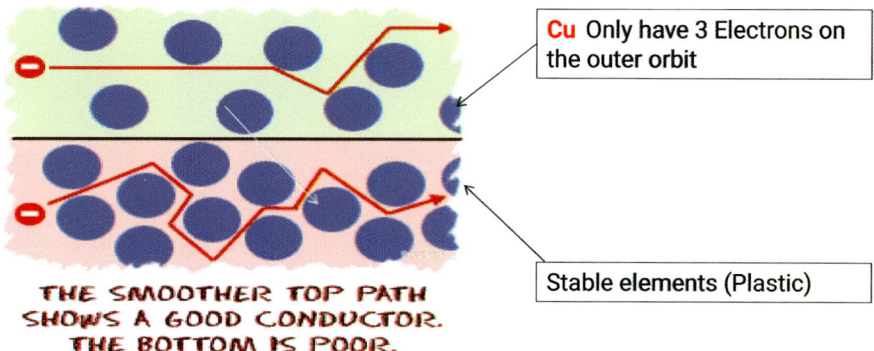

Insulators

Most solid materials are classified as insulators because they offer very large resistance to the flow of electric current.
Some materials are particularly good insulators and can be characterized by their high resistivity:
- Glass: 10^{12} Ohm's of Resistivity
- Mica: 9×10^{13} Ohms of Resistivity
- Quartz (fused) 5×10^{16} Ohms.
- This is compared to the resistivity of copper: **1.7×10^{-8}**

Thermal & Electrical Conductivity

Metal	Thermal Conductivity	Electric Conductivity
Cooper	100	100
Silver	108	105
Iron	17	17
Steel	13-17	15
Brass	28	25
Aluminum	56	62
Zinc	29	30
Lead	9	8

Marine Conductors

Electrons flow from negative post to positive but not until the circuit is completed. The wire should be multi strand so it's flexible and won't snap from fatigue. It should be tinned copper to resist corrosion.

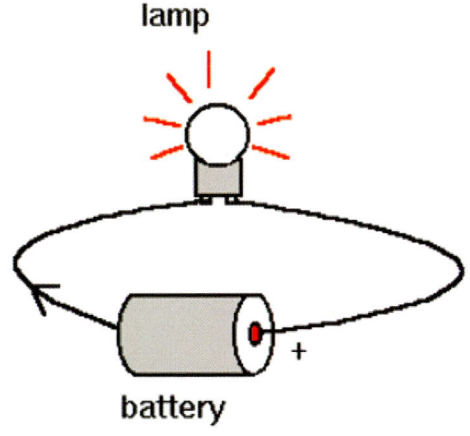

The Box Crimper

For Battery cable gauge sizes, the tool of choice for terminations is referred to as a " Box Crimper" which leaves a square crimp on the terminal.

Electrical Conductors

Electrical wire is usually round in cross-section (although there are some unique exceptions to this rule) and comes in two basic varieties: solid and stranded.

Stranded wire is composed of smaller strands of solid copper wire twisted together to form a single, larger conductor.

The greatest benefit of stranded wire is its mechanical flexibility, being able to withstand repeated bending and twisting much better than solid copper (which tends to fatigue and break after time).

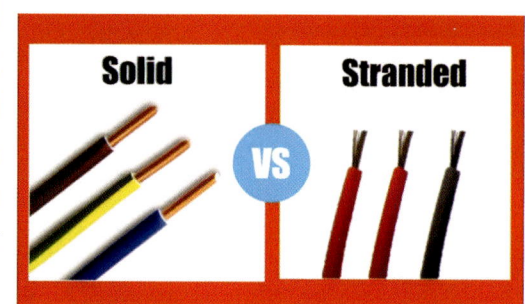

Video Episode 2: Wire gauge calculation & wire color code

Circular –Mil Area

The table below shows that 12 AWG is the correct choice

Size AWG	Diameter Inches	Min Circular Mils
18	.044	1537
16	.055	2336
14	.069	3702
12	.086	5833 5972
10	.109	9343
8	.137	14810
6	.182	25910

In this video, you will learn the procedure to size a wire according to the Amps, Total length, and Voltage drop.

Follow me

Scan this code to see the highlight video

Chapter 4: Conductors and Isolators

Wire Connectors

Most connections are made using crimp terminations
Crimps are color coded to wire gauge.
Crimps should be tinned copper.

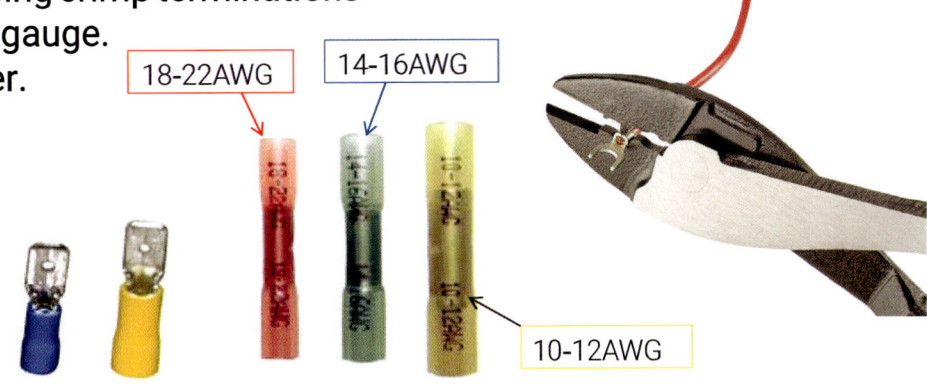

18-22AWG   14-16AWG   10-12AWG

According to ABYC the maximum number of connectors allowed by each terminal is four.

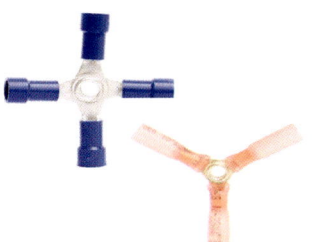

When stripping insulation from wire, use an appropriate tool so that wire strands are not damage.
If they are damaged, cut the wire off and try again.

Chapter 4: Conductors and Isolators

Conductor Size

We could speak of a wire's diameter, but since its really the cross-sectional area that matters most regarding the flow of electrons, we are better off designating wire size in terms of area.

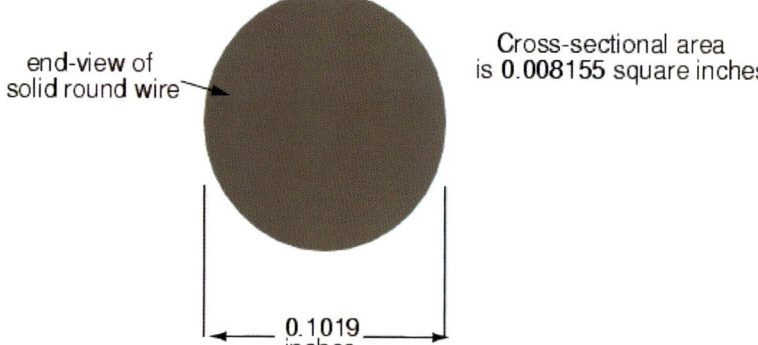

Circular Mills

A circular mil is a unit of area, equal to the area of a circle with a diameter of one mil (one thousandth of an inch or 0.001 in).
The area in circular mils, A, of a circle with a diameter of d mils, is given by the formula:
$A = d^2$ because the area in circular mils can be calculated without reference to pi (π).

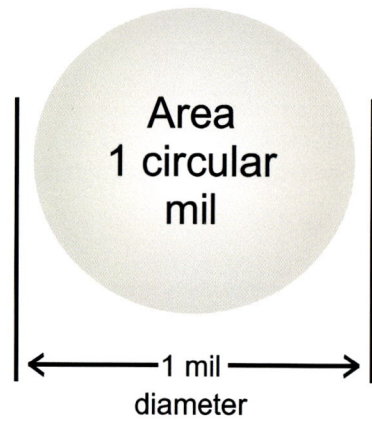

Circular-Mil Area

If a wire has a cross-sectional diameter of 1 mil, by definition, the circular mil area (CMA) is $A = D^2$, or $A = 1^2$, or A = 1 circular mil.
A 12-gauge wire has a diameter of 80.81 mils. What is (1) its area in circular mils?

$$-A = D*D = 6{,}530 \text{ circular mills (CM)}$$

Circular-Mil Area

The table below shows that 12 AWG wire has a CM area of 5,833 and is the correct choice.

Size AWG	Diameter Inches	Min Circular Mils
18	.044	1537
16	.055	2336
14	.069	3702
12	.086	5833
10	.109	9343
8	.137	14810
6	.182	25910

Circular –Mil Area and Stranding

Size AWG	Diameter Inches	Min Circular Mils
4	.218	37360
2	.282	62450
1	.315	77790
0	.355	98980
2/0	.399	125100
3/0	.449	158600
4/0	.512	205500

Voltage Drop

As current flows through the wire, voltage drops according to Ohm's law.
If power is supplied to a circuit by a 12 V battery and the voltage drop 0.5 V in the positive conductor and 0.5 V in the negative back to the battery. Then the voltage across the load is not 12 V is 12 - 0.5 – 0.5 = 11 V
The Voltage drop in the conductor is 1.0/12 = 8.3%

Voltage Drop

The ABYC specifies two allowable percentage drops, depending on the effect on safety
- **3%** for panel boards feeds, bilge blowers, electronics and   navigation lights
- **10%** for general lighting

Conductor size may be determined from
- **CM = (K x I x L)/ E**
- CM = Conductor Circular Mills
- K = 10.75 , I = Current in amps
- L = round-trip length in feet
- E = Voltage drop

Wire Size Calculation

If the voltage drop is of 3% the following formula is recommended by ABYC to calculate the size of the wire.

Area CM = 10.75 x (amps) x (total length of run) / 0.36 (3% of 12v).

For example: A bilge pump draws 10 amps. The positive run is 10 feet from the power panel, including the float switch. The negative run is only 10 feet. What size wire?.
CM = 10.75 x 10 (amps) x 20 (total length of run) / 0.36 (3% of 12v) = 5,972.

Circular –Mil Area

The table below shows that 12 AWG is the correct choice.

Size AWG	Diameter Inches	Min Circular Mils
18	.044	1537
16	.055	2336
14	.069	3702
12	.086	5833
10	.109	9343
8	.137	14810
6	.182	25910

Sizing AWG wire using ABYC Tables (3% Voltage Drop)

Another way to calculate the wire gauge is by using the tables recommended by ABYC. Those tables follow the same criteria. The first two tables are related to a voltage drop of 3% for both 12V and 24V. In the first top row, you will find the total distance from the power source to the load. For calculation purposes, the total distance should be twice the distance between the load and the power source. In the left column, you will find the total current flowing through the circuit in Amps, and in the middle of the table are the wire gauges corresponding to the Distance vs the Amps.

Wire Gauges for a 3-Percent Voltage Drop in a DC 12-Volt System

Length of Conductor from Source of Current to Device and Back to source (feet)

Total Current on Circuit (amps)	10	15	20	25	30	40	50	60	70	80	90	100	110	120	130	140	150	160	170
5	18	16	14	12	12	10	10	10	8	8	8	6	6	6	6	6	6	6	6
10	14	12	10	10	10	8	6	6	6	6	4	4	4	4	2	2	2	2	2
15	12	10	10	8	8	6	6	6	4	4	2	2	2	2	2	1	1	1	1
20	10	10	8	6	6	6	4	4	2	2	2	2	1	1	1	0	0	0	2/0
25	10	8	6	6	6	4	4	2	2	2	1	1	0	0	0	2/0	2/0	2/0	2/0
30	10	8	6	6	4	4	2	2	1	1	0	0	0	2/0	2/0	3/0	3/0	3/0	3/0
40	8	6	6	4	4	2	2	1	0	0	2/0	2/0	3/0	3/0	3/0	4/0	4/0	4/0	4/0
50	6	6	4	4	2	2	1	0	2/0	2/0	3/0	3/0	4/0	4/0	4/0				
60	6	4	4	2	2	1	0	2/0	3/0	3/0	4/0	4/0	4/0						
70	6	4	2	2	1	0	2/0	3/0	3/0	4/0	4/0								
80	6	4	2	2	1	0	3/0	3/0	4/0	4/0									
90	4	2	2	1	0	2/0	3/0	4/0	4/0										
100	4	2	2	1	0	2/0	3/0	4/0	4/0										

Wire Gauges for a 3-Percent Voltage Drop in a DC 24-Volt System

Length of Conductor from Source of Current to Device and Back to source (feet)

Total Current on Circuit (amps)	10	15	20	25	30	40	50	60	70	80	90	100	110	120	130	140	150	160	170
5	18	18	18	16	16	14	12	12	12	10	10	10	10	10	8	8	8	8	8
10	18	16	14	12	12	10	10	10	8	8	8	6	6	6	6	6	6	6	6
15	16	14	12	12	10	10	8	8	6	6	6	6	6	4	4	4	4	4	2
20	14	12	10	10	10	8	6	6	6	6	4	4	4	4	2	2	2	2	2
25	12	12	10	10	8	6	6	6	4	4	4	4	2	2	2	2	2	2	2
30	12	10	10	8	8	6	6	4	4	4	2	2	2	2	2	1	1	1	1
40	10	10	8	6	6	4	4	2	2	2	2	2	1	1	1	0	0	0	2/0
50	10	8	6	6	6	4	4	2	2	2	1	1	0	0	0	2/0	2/0	2/0	3/0
60	10	8	6	6	4	4	2	2	1	1	0	0	0	2/0	2/0	3/0	3/0	3/0	3/0
70	8	6	6	4	4	2	2	1	1	0	0	2/0	2/0	3/0	3/0	3/0	3/0	4/0	4/0
80	8	6	6	4	4	2	2	1	0	0	2/0	2/0	3/0	3/0	3/0	4/0	4/0	4/0	4/0
90	8	6	4	4	2	2	1	0	0	2/0	2/0	3/0	3/0	4/0	4/0	4/0	4/0	4/0	
100	6	6	4	4	2	2	1	0	2/0	2/0	2/0	3/0	4/0	4/0	4/0				

As you look at these figures, keep in mind that higher wire-gauge numbers mean smaller wire diameters. There's an inverse relationship.

AC and DC Circuits With Bundled and Sheathed Conductors

For AC and DC circuits depending if the current carrying conductors are bundled or unbundled and sheathed or unsheathed the following tables are used to calculate the wire gauge including the temperature rating in the criteria to select the proper gauge.

Conductor Size (AWG)	Temperature Rating of Conductor Insulation													
	60°C (140°F)		75°C (167°F)		80°C (176°F)		90°C (194°F)		105°C (221°F)		125°C (257°F)		200°C (392°F)
	Outside Engine Spaces	Inside Engine spaces	Outside Engine Spaces	Inside Engine spaces	Outside Engine Spaces	Inside Engine spaces	Outside Engine Spaces	Inside Engine spaces	Outside Engine Spaces	Inside Engine spaces	Outside Engine Spaces	Inside Engine spaces	Outside or Inside Engine spaces
18	10	NOT PERMITTED	10	7.5	15	11.7	20	16.4	20	17.0	25	22.3	25
16	15		15	11.3	20	15.6	25	20.5	25	21.3	30	26.7	35
14	20		20	15.0	25	19.5	30	24.6	35	29.8	40	35.6	45
12	25		25	18.8	35	27.3	40	32.8	45	38.3	50	44.5	55
10	40		40	30.0	50	39.0	55	45.1	60	51.0	70	62.3	70
8	55		65	48.8	70	54.6	70	57.4	80	68.0	90	80.1	100
6	80		95	71.3	100	78.0	100	82.0	120	102.0	125	111.3	135
4	105		125	93.0	130	101.4	135	110.7	160	136.0	170	151.3	180
3	120		145	108.8	150	117.0	155	127.1	180	153.0	195	173.6	210
2	140		170	127.5	175	136.5	180	147.6	210	178.5	225	200.3	240
1	165		195	146.3	210	163.8	210	172.2	245	208.3	265	235.9	280
0	195		230	172.5	245	191.1	245	200.9	285	242.3	305	271.5	325
00	225		265	198.8	285	222.3	285	233.7	330	280.5	355	316.0	370
000	260		310	232.5	330	257.4	330	270.6	385	327.3	410	364.9	430
0000	300		360	270.0	385	300.3	385	315.7	445	378.3	475	422.8	510

AC & DC circuits: Allowable amperage of single conductors not bundled, sheathed, or in conduit.
This reproduces Table VI-A from Standard E-11.

Sizing AWG wire using ABYC Tables (10% Voltage Drop)

The second pair of tables are related to a voltage drop of 10%.

Wire Gauges for a 10-Percent Voltage Drop in a DC 12-Volt System

Total Current on Circuit (amps)	Length of Conductor from Source of Current to Device and Back to source (feet)																	
		10	15	20	25	30	40	50	60	70	80	90	100	110	120	130	140	150	160	170
5	18	18	18	18	18	16	16	14	14	14	12	12	12	12	12	10	10	10	10
10	18	18	16	16	14	14	12	12	10	10	10	10	8	8	8	8	8	8	6
15	18	16	14	14	12	12	10	10	8	8	8	8	8	6	6	6	6	6	6
20	16	14	14	12	12	10	10	8	8	8	6	6	6	6	6	6	4	4	4
25	16	14	12	12	10	10	8	8	6	6	6	6	6	4	4	4	4	4	2
30	14	12	12	10	10	8	8	6	6	6	6	4	4	4	4	2	2	2	2
40	14	12	10	10	8	8	6	6	6	4	4	4	2	2	2	2	2	2	2
50	12	10	10	8	8	6	6	4	4	4	2	2	2	2	2	1	1	1	1
60	12	10	8	8	6	6	4	4	2	2	2	2	2	1	1	1	0	0	0
70	10	8	8	6	6	6	4	2	2	2	2	1	1	1	0	0	0	2/0	2/0
80	10	8	8	6	6	4	4	2	2	2	1	1	0	0	0	2/0	2/0	2/0	2/0
90	10	8	6	6	6	4	2	2	2	1	1	0	0	0	2/0	2/0	2/0	3/0	3/0
100	10	8	6	6	4	4	2	2	1	1	0	0	0	2/0	2/0	2/0	3/0	3/0	3/0

Wire Gauges for a 10-Percent Voltage Drop in a DC 24-Volt System

Total Current on Circuit (amps)	Length of Conductor from Source of Current to Device and Back to source (feet)																	
		10	15	20	25	30	40	50	60	70	80	90	100	110	120	130	140	150	160	170
5	18	18	18	18	18	18	18	18	16	16	16	16	14	14	14	14	14	14	12
10	18	18	18	18	18	16	16	14	14	14	12	12	12	12	12	10	10	10	10
15	18	18	18	16	16	14	14	12	12	12	10	10	10	10	10	8	8	8	8
20	18	18	16	16	14	14	12	12	10	10	10	8	8	8	8	8	8	8	6
25	18	16	16	14	14	12	12	10	10	10	8	8	8	8	8	6	6	6	6
30	18	16	14	14	12	12	10	10	8	8	8	8	8	6	6	6	6	6	6
40	16	14	14	12	12	10	10	8	8	8	6	6	6	6	6	6	4	4	4
50	16	14	12	12	10	10	8	8	6	6	6	6	6	4	4	4	4	4	2
60	14	12	12	10	10	8	8	6	6	6	6	4	4	4	4	2	2	2	2
70	14	12	10	10	8	8	6	6	6	6	4	4	4	2	2	2	2	2	2
80	14	12	10	10	8	8	6	6	6	4	4	4	2	2	2	2	2	2	2
90	12	10	10	8	8	6	6	6	4	4	4	2	2	2	2	2	2	1	1
100	12	10	10	8	8	6	6	4	4	4	2	2	2	2	2	1	1	1	1

Conductors introduce electrical resistance, and some voltage drop is enevitable.
As amperage and length increase, so must the wire size.

Chapter 4: Conductors and Isolators

Conductor Size (AWG)	Temperature Rating of Conductor Insulation													
	60°C (140°F)		75°C (167°F)		80°C (176°F)		90°C (194°F)		105°C (221°F)		125°C (257°F)		200°C (392°F)
	Outside Engine Spaces	Inside Engine spaces	Outside Engine Spaces	Inside Engine spaces	Outside Engine Spaces	Inside Engine spaces	Outside Engine Spaces	Inside Engine spaces	Outside Engine Spaces	Inside Engine spaces	Outside Engine Spaces	Inside Engine spaces	Outside or Inside Engine spaces
18	7.0	NOT PERMITTED	7.0	5.3	10.5	8.2	14.0	11.5	14.0	11.9	17.5	15.6	17.5
16	10.5		10.5	7.9	14.0	10.9	17.5	14.4	17.5	14.9	21.0	18.7	24.5
14	14.0		14.0	10.5	17.5	13.7	21.0	17.2	24.5	20.8	28.0	24.9	31.5
12	17.5		17.5	13.1	24.5	19.1	28.0	23.0	31.5	26.8	35.0	31.5	38.5
10	28.0		28.0	21.0	35.0	27.3	38.5	31.6	42.0	35.7	49.0	43.6	49.0
8	38.5		45.5	34.1	49.0	38.2	49.0	40.2	56.0	47.6	63.0	56.1	70.0
6	56.0		66.5	49.9	70.0	54.6	70.0	57.4	84.0	71.4	87.5	77.9	94.5
4	73.5		87.5	65.6	91.0	71.0	94.5	77.5	112.0	95.5	119.0	105.9	126.0
3	84.0		101.5	76.1	105.0	81.9	108.5	89.0	126.0	107.1	136.5	121.5	147.0
2	98.0		119.0	89.3	122.5	95.6	126.0	103.3	147.0	125.0	157.5	140.2	168.0
1	115.5		136.5	102.4	147.0	114.7	147.0	120.5	171.5	145.8	185.5	165.1	196.0
0	136.5		161.0	120.8	171.0	133.8	171.5	140.6	199.5	169.6	213.5	190.0	227.5
00	157.5		185.5	139.1	199.5	155.6	199.5	163.6	231.0	196.4	248.5	221.2	259.0
000	182.0		217.0	162.8	231.0	180.2	231.0	189.4	269.5	229.1	287.0	255.4	301.0
0000	210.0		252.0	189.0	269.5	210.2	269.5	221.0	311.5	264.8	332.5	295.9	357.0

AC & DC circuits: **Allowable** amperage of conductors when up to three current-carrying conductors are bundles, sheathed or in conduit. This reproduces the values from Standard E-11, Table VI-B

Conductor Size (AWG)	Temperature Rating of Conductor Insulation													
	60°C (140°F)		75°C (167°F)		80°C (176°F)		90°C (194°F)		105°C (221°F)		125°C (257°F)		200°C (392°F)
	Outside Engine Spaces	Inside Engine spaces	Outside Engine Spaces	Inside Engine spaces	Outside Engine Spaces	Inside Engine spaces	Outside Engine Spaces	Inside Engine spaces	Outside Engine Spaces	Inside Engine spaces	Outside Engine Spaces	Inside Engine spaces	Outside or Inside Engine spaces
18	6.0	NOT PERMITTED	6.0	4.5	9.0	7.0	12.0	9.8	12.0	10.2	15.0	13.4	15.0
16	9.0		9.0	6.8	12.0	9.4	15.0	12.3	15.0	12.8	18.0	16.0	21.0
14	12.0		12.0	9.0	15.0	11.7	18.0	14.8	21.0	17.9	24.0	21.4	27.0
12	15.0		15.0	11.3	21.0	16.4	24.0	19.7	27.0	23.0	30.0	26.7	33.0
10	24.0		24.0	18.0	30.0	23.4	33.0	27.1	36.0	30.6	42.0	37.4	42.0
8	33.0		39.0	29.3	42.0	32.8	42.0	34.4	48.0	40.8	54.0	48.1	60.0
6	48.0		57.0	42.8	60.0	46.8	60.0	49.2	72.0	61.2	75.0	66.8	81.0
4	63.0		75.0	56.3	78.0	60.8	81.0	66.4	96.0	81.6	102.0	90.8	108.0
3	72.0		87.0	65.3	90.0	70.2	93.0	76.6	108.0	91.8	117.0	104.1	126.0
2	84.0		102.0	76.5	105.0	81.9	108.0	88.6	126.0	107.1	15.0	120.2	144.0
1	99.0		117.0	87.8	126.0	98.3	126.0	103.3	147.0	125.0	159.0	141.5	168.0
0	117.0		138.0	103.5	147.0	114.7	147.0	120.5	171.0	145.4	183.0	162.9	195.0
00	135.0		159.0	119.3	171.0	133.4	171.0	140.2	198.0	168.3	214.0	189.6	222.0
000	156.0		186.0	139.5	198.0	154.4	198.0	162.4	231.0	196.4	246.0	218.9	258.0
0000	180.0		216.0	162.0	231.0	180.2	231.0	189.4	267.0	227.0	285.0	253.7	306.0

AC circuits: Allowable amperage of conductors when four to six current-carrying conductors are bundles. Table VI-C from Standard E-11.

Chapter 4: Conductors and Isolators

| Conductor Size (AWG) | Temperature Rating of Conductor Insulation |||||||||||||
| | 60°C (140°F) || 75°C (167°F) || 80°C (176°F) || 90°C (194°F) || 105°C (221°F) || 125°C (257°F) || 200°C (392°F) |
	Outside Engine Spaces	Inside Engine spaces	Outside Engine Spaces	Inside Engine spaces	Outside Engine Spaces	Inside Engine spaces	Outside Engine Spaces	Inside Engine spaces	Outside Engine Spaces	Inside Engine spaces	Outside Engine Spaces	Inside Engine spaces	Outside or Inside Engine spaces
18	5.0	NOT PERMITTED	5.0	3.8	7.5	5.9	10.0	8.2	10.0	8.5	12.5	11.1	12.5
16	7.5	NOT PERMITTED	7.5	5.6	10.0	7.8	12.5	10.3	12.5	10.6	15.0	13.4	17.5
14	10.0	NOT PERMITTED	10.0	7.5	12.5	9.8	15.0	12.3	17.5	14.9	20.0	17.8	22.5
12	12.5	NOT PERMITTED	12.5	9.4	17.5	13.7	20.0	16.4	22.5	19.1	25.0	22.3	27.5
10	20.0	NOT PERMITTED	20.0	15.0	25.0	19.5	27.5	22.6	30.0	25.5	35.0	31.2	35.0
8	27.5	NOT PERMITTED	32.5	24.4	35.0	27.3	35.0	28.7	40.0	34.0	45.0	40.1	50.0
6	40.0	NOT PERMITTED	47.5	35.6	50.0	39.0	50.0	41.0	60.0	51.0	62.5	55.6	67.5
4	52.5	NOT PERMITTED	62.5	46.9	65.0	50.7	67.5	55.4	80.0	68.0	85.0	75.7	90.0
3	60.0	NOT PERMITTED	72.5	54.4	75.0	58.5	77.5	63.6	90.0	76.5	97.5	86.8	105.0
2	70.0	NOT PERMITTED	85.0	63.8	87.5	68.3	90.0	73.8	105.0	89.3	112.5	100.1	120.0
1	82.5	NOT PERMITTED	97.5	73.1	105.0	81.9	105.0	86.1	122.5	104.1	132.5	117.9	140.0
0	97.5	NOT PERMITTED	115.0	86.3	122.5	95.6	122.5	100.5	142.5	121.1	152.5	135.7	162.5
00	112.5	NOT PERMITTED	132.5	99.4	142.5	111.2	142.5	116.9	165.0	140.3	177.5	158.0	185.0
000	130.0	NOT PERMITTED	155.0	116.3	165.0	128.7	165.0	135.3	192.5	163.6	205.0	182.5	215.0
0000	150.0	NOT PERMITTED	180.0	135.0	192.5	150.2	192.5	157.9	222.5	189.1	237.5	211.4	255.0

AC circuits: Allowable amperage of conductors when 7 to 24 current-carrying conductors are bundles.
Table VI-D from Standard E-11.

| Conductor Size (AWG) | Temperature Rating of Conductor Insulation |||||||||||||
| | 60°C (140°F) || 75°C (167°F) || 80°C (176°F) || 90°C (194°F) || 105°C (221°F) || 125°C (257°F) || 200°C (392°F) |
	Outside Engine Spaces	Inside Engine spaces	Outside Engine Spaces	Inside Engine spaces	Outside Engine Spaces	Inside Engine spaces	Outside Engine Spaces	Inside Engine spaces	Outside Engine Spaces	Inside Engine spaces	Outside Engine Spaces	Inside Engine spaces	Outside or Inside Engine spaces
18	4.0	NOT PERMITTED	4.0	3.0	6.0	4.7	8.0	6.6	8.0	6.8	10.0	8.9	10.0
16	6.0	NOT PERMITTED	6.0	4.5	8.0	6.2	10.0	8.2	10.0	8.5	12.0	10.7	14.0
14	8.0	NOT PERMITTED	8.0	6.0	10.0	7.8	12.0	9.8	14.0	11.9	16.0	14.2	18.0
12	10.0	NOT PERMITTED	10.0	7.5	14.0	10.9	16.0	13.1	18.0	15.3	20.0	17.8	22.0
10	16.0	NOT PERMITTED	16.0	12.0	20.0	15.6	22.0	18.0	24.0	20.4	28.0	24.9	28.0
8	22.0	NOT PERMITTED	26.0	19.5	28.0	21.8	27.0	23.0	32.0	27.2	36.0	32.0	40.0
6	32.0	NOT PERMITTED	38.0	28.5	40.0	31.2	40.0	32.8	48.0	40.8	50.0	44.5	54.0
4	42.0	NOT PERMITTED	50.0	37.5	52.0	40.6	54.0	44.3	64.0	54.4	68.0	60.5	72.0
3	48.0	NOT PERMITTED	58.0	43.5	60.0	46.8	62.0	50.8	72.0	61.2	78.0	69.4	84.0
2	56.0	NOT PERMITTED	68.0	51.0	70.0	54.6	72.0	59.0	84.0	71.4	90.0	80.1	96.0
1	66.0	NOT PERMITTED	78.0	58.5	84.0	65.5	84.0	68.9	98.0	83.3	106.0	94.3	112.0
0	78.0	NOT PERMITTED	92.0	69.0	98.0	76.4	98.0	80.4	114.0	96.9	122.0	108.6	130.0
00	90.0	NOT PERMITTED	106.0	79.5	114.0	88.9	114.0	93.5	132.0	112.2	142.0	126.4	148.0
000	104.0	NOT PERMITTED	124.0	93.0	132.0	103.0	132.0	108.2	154.0	130.9	164.0	146.0	172.0
0000	120.0	NOT PERMITTED	144.0	108.0	154.0	120.1	154.0	126.3	178.0	151.3	190.0	169.1	204.0

AC circuits: Allowable amperage of conductors when 25 or more current-carrying conductors are bundled.
Table VI-E from Standard E-11.

Marine Wire Gauges

ABYC calls for all conductors to be stranded and of size <u>AWG 16 minimum.</u>
The single exception is sheathed AWG 18 conductors that do not extend more than 30 inches beyond their sheath.
Conductor size should be selected considering conductor length and current in order the voltage drop not impair the functioning of the loads.

Marine Grade Wire

Marine grade wire is available as either boat cable or in spools of stranded insulated wire. Boat cable consists of either two or three insulated stranded wires within a white vinyl outer wrap, and is clearly marked as marine boat cable.

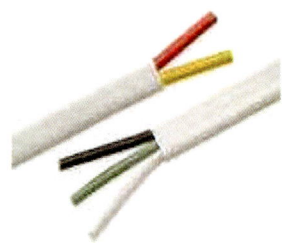  

Spools of wire are available with insulation colored to match standard marine wiring standards and are also clearly marked as meeting specifications for marine use.
Insist on Marine Grade boat cable. It is UL approved for the corrosive marine environment. It is designed to exceed all test standards for cold bend, moisture and oil resistance, heat shock and flammability.

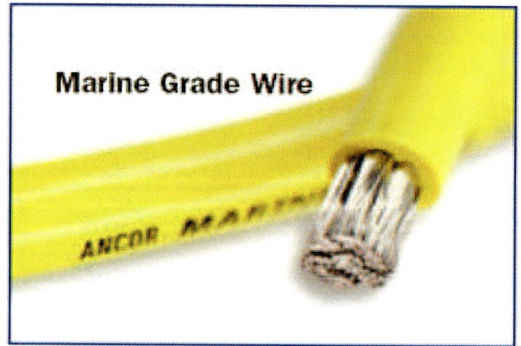

Colors for DC Power

Standard colors as approved by the American Boat and Yacht Council are **red** for DC positive conductors, **black** or **yellow** for DC negative, and green or **green** with a yellow stripe for DC grounding.

ABYC revised the marine electrical standards which started to show up in some boats as early as the 1996 model year. A big change is that the 12 volt black ground wire color was changed to making **"Yellow"** an optional color as a negative or ground wire.

Electrical Wire Gauge and Connectors

Marine Wire Color Code DC Current

COLOR	USE
Green	Bonding System
Yellow or Black	Ground
Red	Main Power feeds
Yellow w/ Red	Starting Systems
Brown w/ Yellow	Bilge Blowers
Dark Gray	Nav Lights / Tachometer
Brown	Pumps, Alternator Charge light
Orange	Accessory Feed, Common Feed
Purple	Ignition, Instrument Feed
Dark Blue	Cabin & Instruments
Light Blue	Oil Pressure
Tan	Water Temperature
Pink	Fuel Gauge
Brown	Pumps, Generator Armature

Wire Sizes

Wire size is as important as everything else. Always use the proper size wire for the electrical load it will be carrying.
For circuits which are relatively short use a wire gauge one size larger than the size wire the accessory you are connecting provides.

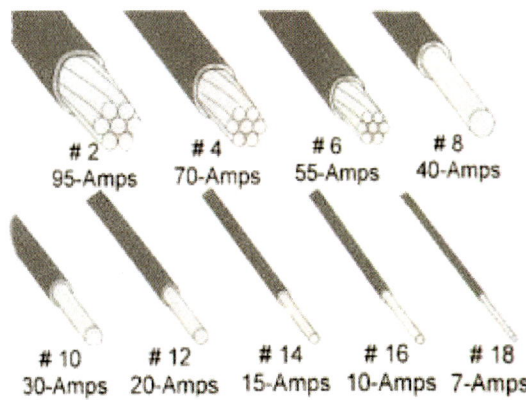

Types of Marine Wire

The smaller AWG (American Wire Gauge) sizes (16 to 8 AWG) of single conductor wire are called primary wires.

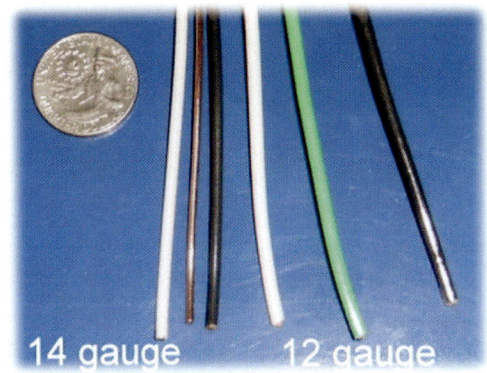

These usually make up the circuit wiring of a boat's auxiliary systems to the main battery. Typically used for DC electrical switches & controls.

Large (6 to 4/0 AWG), single conductor wire is referred to as battery cable.
The cable usually has a high tin plated copper strand count. The tin plated copper helps minimize corrosion and extend the life of the cable. Typically used for battery installation and DC main power take offs.

Marine Duplex Wire

These can help simplify wire installation by reducing the number of separate wires run throughout a boat.
Different color codes are used depending on the application. AC wiring uses **back** & white. DC uses **black** & **red** or **red** & **yellow**. **Red** & **yellow** is the safety color to eliminate the risk of connecting black AC with black DC.

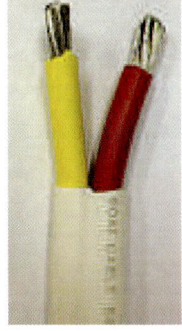

Colors for AC 120V / 60 Hz American

Black to AC "HOT"
White to AC "NEUTRAL"
Green to AC "GROUND"

American 60Hz Black, White & Green

Colors for AC 240 V / 60 Hz American

Black to AC "HOT 1"
Red to AC "HOT 2".
White to AC "NEUTRAL"
Green to AC "GROUND"

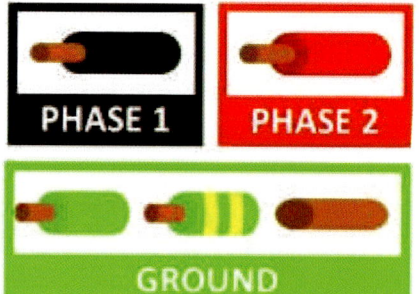

Colors for AC 230V / 50 Hz European

Brown to AC "HOT ".
Blue to AC "NEUTRAL"
Green/yellow to AC "GROUND"

European 50 Hz Brown, Blue & Green/Yellow

Colors for AC 460 V / 50 Hz European

Brown to AC "HOT 1"
Black to AC "HOT 2"
Blue to AC "NEUTRAL"
Green/yellow to AC "GROUND"

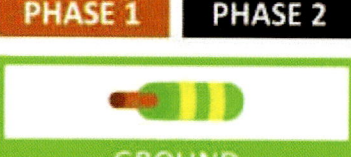

Colors for DC 12V / 24V

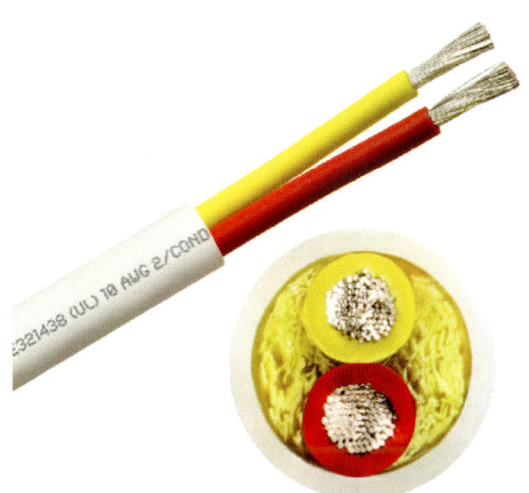

Yellow wire is recommended as DC negative because black is the standard color for AC hot. There have been many cases of people working on their DC systems who have inadvertently cut the live AC wire.

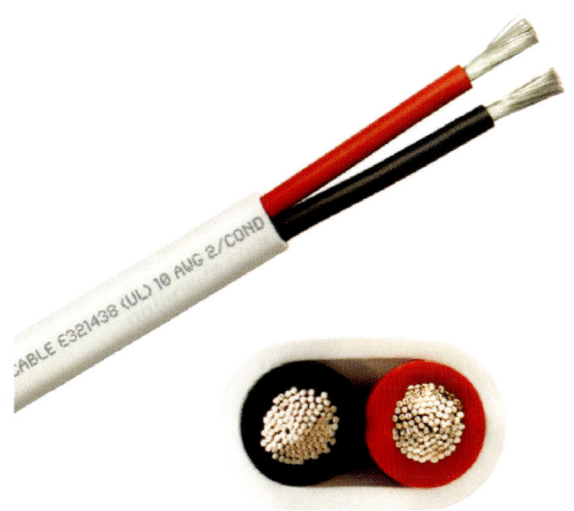

Wire Sizes

Sizes vary according to the application, with lower numbers indicating the ability to carry heavier current loads.
The bigger the number the lower the current flow capacity
For example, #4 AWG is commonly used for high amperage battery connections and #14 AWG for "house" wiring.

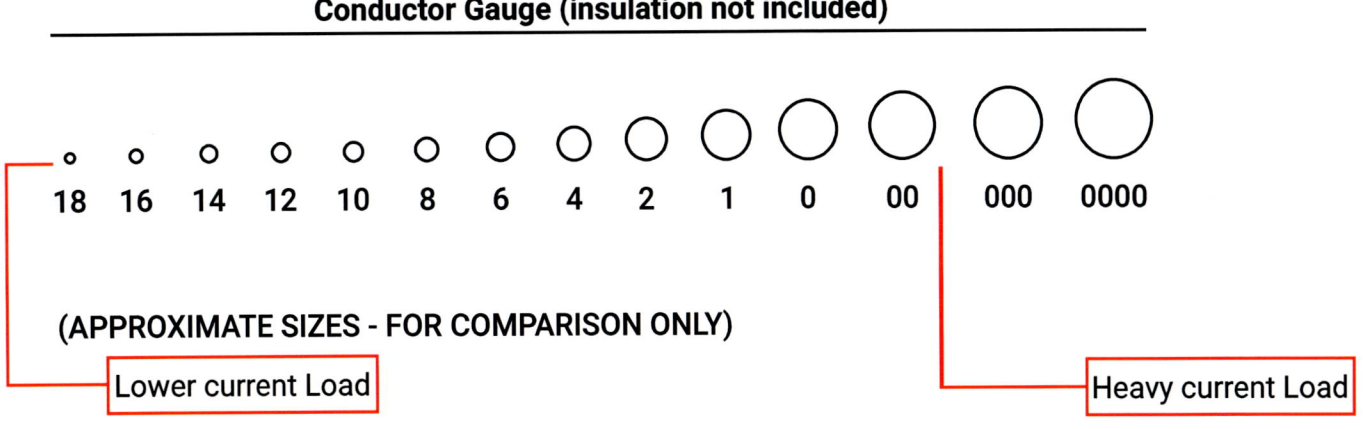

Marine Triplex Wire

Another common multi-conductor boat cable, triplex wire can also reduce time spent in installation. The usual color code for triplex wire is **black**, white & green (designed for AC as hot, neutral & ground). Typically used for AC wiring.

Checking for Continuity

With the advent of modern equipment, the ohmmeter is probably not the best choice today. The (TDR) Time Domain Reflectometer is a good tool to confirming continuity and location of short and open circuits.

Abandoning Old Wiring

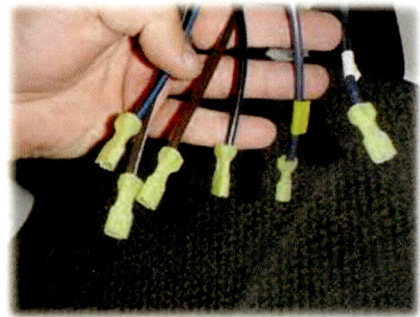

Remove old, unused wiring is not always practical or cost-effective.
The left-behind cabling should have its ends insulated, labeled as to their status and folded back out of the way of functional wiring.

Video Episode 3: Electrical Measurement Tools

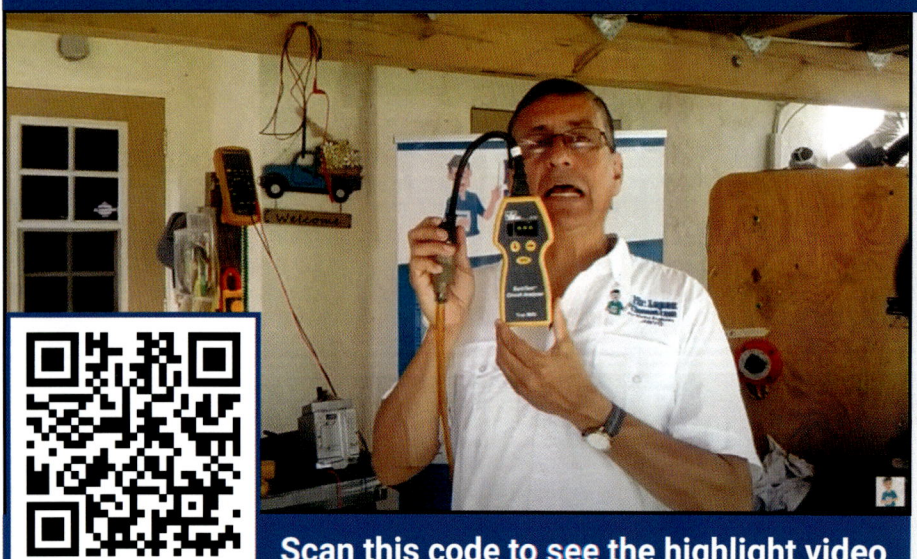

Scan this code to see the highlight video

In this video you will find all the electrical measurement tools commonly used in marine installation and reparations. This is a great video for boat owner and marine technicians interested to develop a career in the marine field.

Follow me

CHAPTER 5

Types of Circuits

TOPICS

5.1 DC Circuits	60
5.2 Open / Short Circuits	60
5.3 Series Circuits	61
5.4 Parallel Circuits	63
5.5 Combined Circuits	66
5.6 Power Calculation	71

Video Episode 1: Solving electrical circuits Series and Parallel

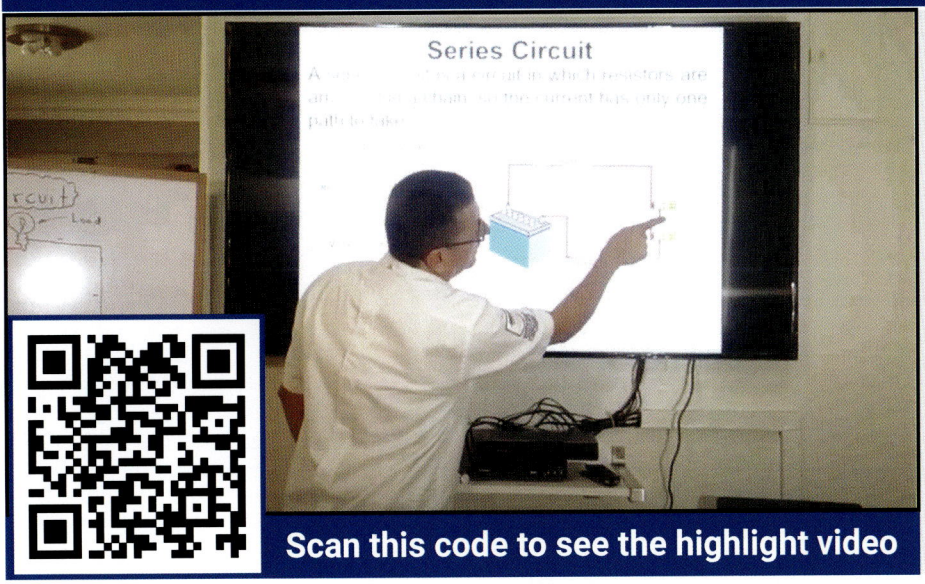

Scan this code to see the highlight video

In this video we are going to learn the fundamentals of electricity applied for boat construction.

Follow me

Chapter 5: Types of Circuits

DC Circuit

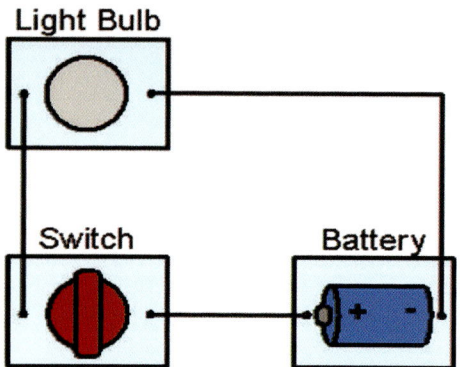

A Direct Current (**DC**) circuit is a circuit that Electric Current flows in one direction.
In marine applications, **DC** circuits are commonly found in 12V, 24V, and 36V, feeding from specific battery banks.
Remember all battery banks, doesn't matter their capacity in Amps and their voltage should be connected together into a common negative bus bar (Only the negative terminals).

Open Circuits

An electric circuit in which the normal path of current has been interrupted is called an open circuit.

Short Circuits

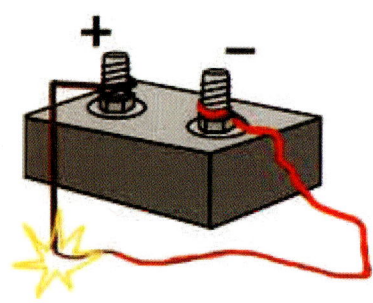

A low-resistance connection established by accident or intention between two points in an electric circuit. The current tends to flow through the area of low resistance, bypassing the rest of the circuit.

Shorts to Ground

A short-circuit to ground meaning that the wire carrying the signal or power is now touching the wire connected to ground.

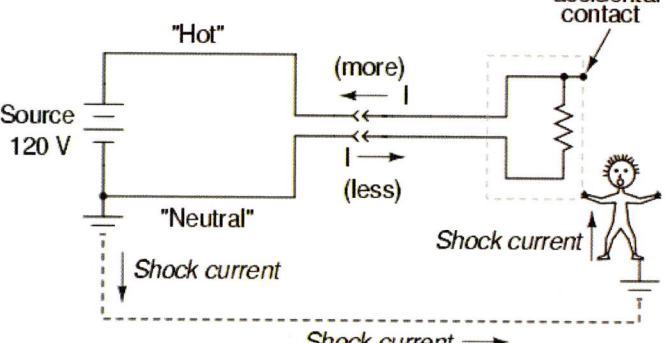

Series Circuit

A series circuit is a circuit in which resistors are arranged in a chain, so the current has only one path to take.

I Total = I1 = I2 = I3 = Kte

Rtotal = R1 + R2 + R3 . . . . . .

V total = V1 + V2 + V3 + ...

The current is the same through each resistor. The total resistance of the circuit is found by simply adding up the resistance values of the individual resistors.

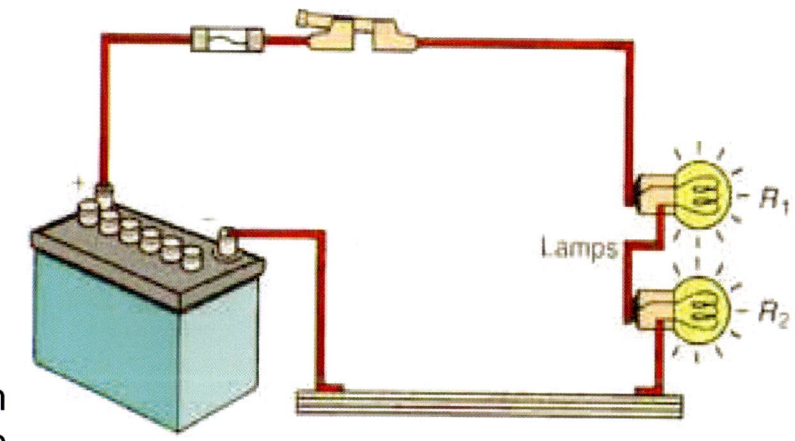

Rtotal = R1 + R2 + R3 . . . . . .

V total = V1 + V2 + V3 + ...

I Total = I1 = I2 = I3 = Kte

Solving Series Circuits

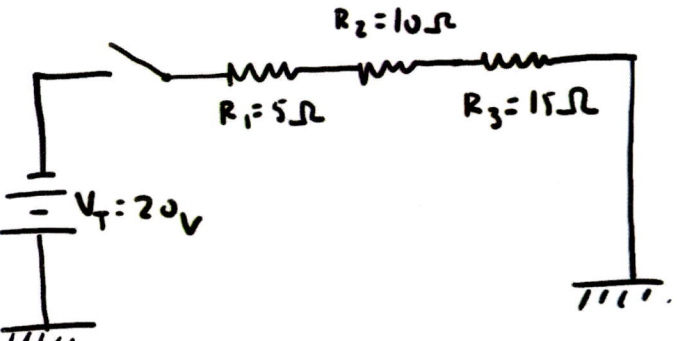

	$V_T = 20$	V_1	V_2	$V_3 =$
	$I_T = \frac{2}{3}$	I_1	$I_2 =$	$I_3 =$
	R_T	$R_1 = 5$	$R_2 = 10$	$R_3 = 15$

Series Circuit $V_T = V_1 + V_2 + V_3 \cdots$
$$I_T = I_1 = I_2 = I_3 = \cdots \text{ cte}$$

1st Step $\Rightarrow$ $R_T = R_1 + R_2 + R_3 \Rightarrow R_T = 5 + 10 + 15 = 30\Omega = R_T$

Solving Series Circuits

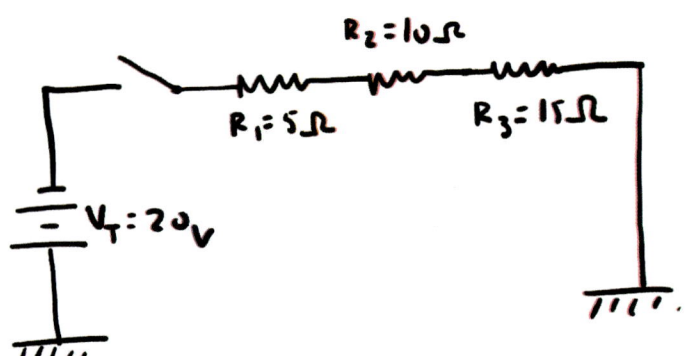

Series Circuit $V_T = V_1 + V_2 + V_3 \ldots$

$\boxed{I_T = I_1 = I_2 = I_3 = \ldots \text{cte}}$

1st Step $\Rightarrow$ $R_T = R_1 + R_2 + R_3 \Rightarrow R_T = 5 + 10 + 15 = 30\,\Omega = R_T$

2nd Step $V_T = ?$ $\Rightarrow$ If $R_T = 30\,\Omega$ and $V_T = 20_V \Rightarrow I_T = \dfrac{20}{30} = \dfrac{2}{3}\,A$

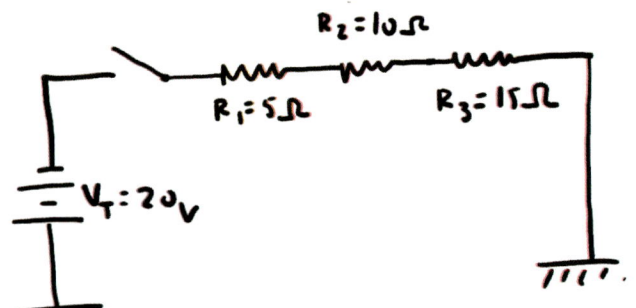

Series Circuit $V_T = V_1 + V_2 + V_3 \ldots$

$\boxed{I_T = I_1 = I_2 = I_3 = \ldots \text{cte}}$

1st Step $\Rightarrow$ $R_T = R_1 + R_2 + R_3 \Rightarrow R_T = 5 + 10 + 15 = 30\,\Omega = R_T$

2nd Step $V_T = ?$ $\Rightarrow$ If $R_T = 30\,\Omega$ and $V_T = 20_V \Rightarrow I_T = \dfrac{20}{30} = \dfrac{2}{3}\,A$

3rd Step $V_1 = ?$ $\Rightarrow$ $V_1 = \dfrac{2}{3}A \times 5\Omega = \dfrac{10}{3}\,V \quad V_2 = \dfrac{2}{3} \times 10 = \dfrac{20}{3} \quad V_3 = \dfrac{30}{3}$

$V_T = 20 = \dfrac{10}{3} + \dfrac{20}{3} + \dfrac{30}{3} = \dfrac{60}{3} = 20_V$ ✓

Mr Lopez.

Chapter 5: Types of Circuits

Solving Series Circuits

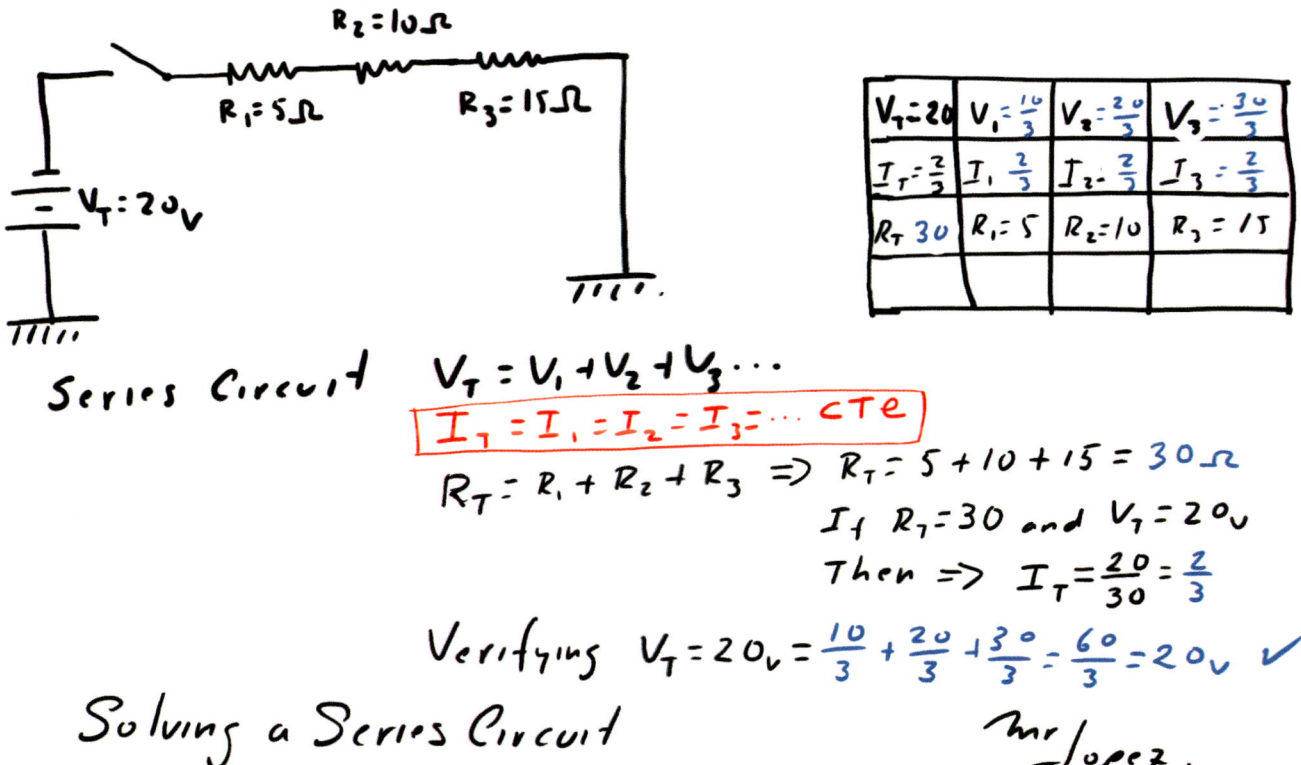

Series Circuit $\quad V_T = V_1 + V_2 + V_3 \ldots$

$\boxed{I_T = I_1 = I_2 = I_3 = \ldots cte}$

$R_T = R_1 + R_2 + R_3 \Rightarrow R_T = 5 + 10 + 15 = 30\,\Omega$

If $R_T = 30$ and $V_T = 20_v$

Then $\Rightarrow I_T = \frac{20}{30} = \frac{2}{3}$

Verifying $V_T = 20_v = \frac{10}{3} + \frac{20}{3} + \frac{30}{3} = \frac{60}{3} = 20_v$ ✓

Solving a Series Circuit

Mr. Lopez.

Parallel Circuit

A parallel circuit is a circuit in which the resistors are arranged with their heads connected together, and their tails connected together.

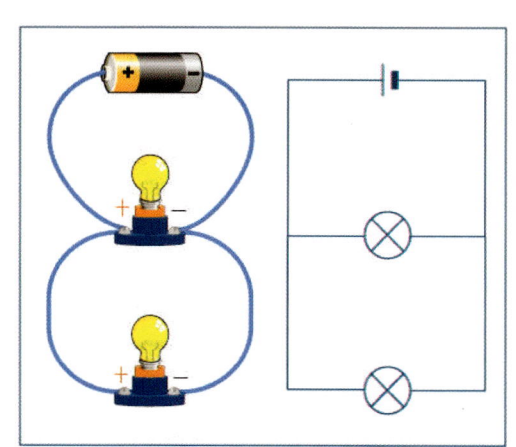

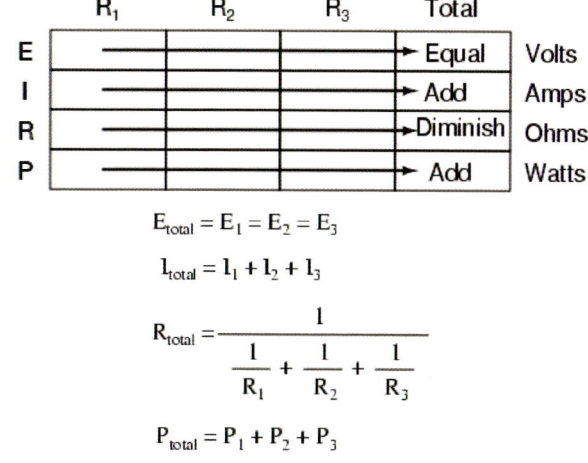

For parallel circuits:

	R_1	R_2	R_3	Total	
E				→ Equal	Volts
I				→ Add	Amps
R				→ Diminish	Ohms
P				→ Add	Watts

$E_{total} = E_1 = E_2 = E_3$

$I_{total} = I_1 + I_2 + I_3$

$R_{total} = \dfrac{1}{\dfrac{1}{R_1} + \dfrac{1}{R_2} + \dfrac{1}{R_3}}$

$P_{total} = P_1 + P_2 + P_3$

Parallel Circuit

The current in a parallel circuit breaks up, with some flowing along each parallel branch and re-combining when the branches meet again. **The voltage across each resistor in parallel is the same.**

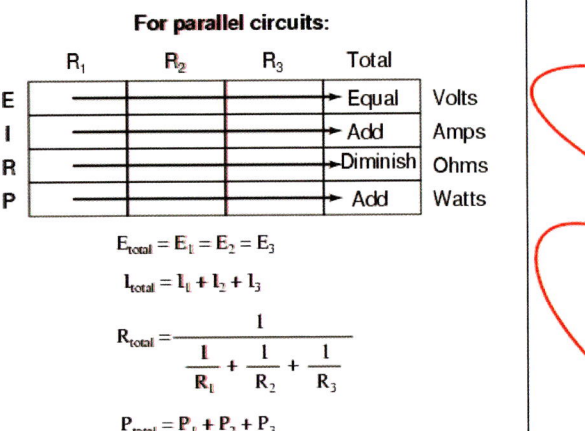

The total resistance of a set of resistors in parallel is found by adding up the reciprocals of the resistance values, and then taking the reciprocal of the total:
The total resistance is: **1 / R = 1 / R1 + 1 / R2 + 1 / R3 +...**

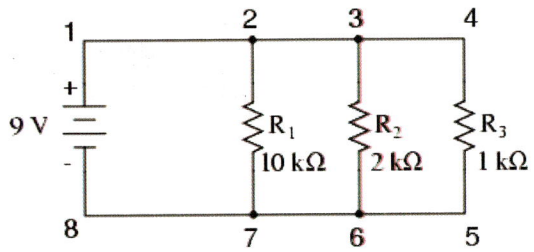

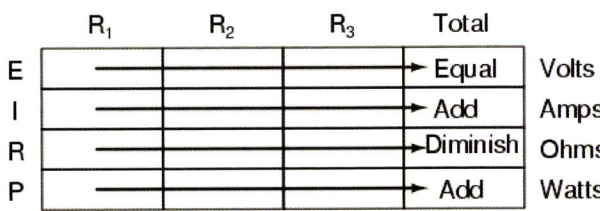

Circuits Series and Parallel

Chapter 5: Types of Circuits

Solving a Parallel Circuit

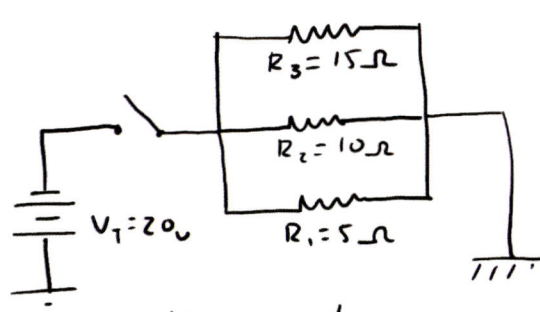

$V_T = 20$	$V_1 = 20$	$V_2 = 20$	$V_3 = 20$
$I_T \frac{22}{3}$	$I_1 = 4$	$I_2 = 2$	$I_3 = \frac{4}{3}$
$R_T = 2.7$	$R_1 = 5$	$R_2 = 10$	$R_3 = 15$
P_T			

For Parallel Circuits

$\boxed{V_1 = V_2 = V_3 = \cdots \text{cte}}$

1st Step $R_T = ?$

Due to a Parallel Circuit $\boxed{V_T = 20 = V_1 = V_2 = V_3 = 20}$

$\dfrac{1}{R_T} = \dfrac{1}{5} + \dfrac{1}{10} + \dfrac{1}{15} = \dfrac{6+3+2}{30} = \dfrac{11}{30} \Rightarrow R_T = \dfrac{30}{11} = 2.72$

Total Resistance is less than the smallest Resistor

$R_T = 2.7 < 5\Omega$ ✓

$V_T = 20$	$V_1 = 20$	$V_2 = 20$	$V_3 = 20$
$I_T \frac{22}{3}$	$I_1 = 4$	$I_2 = 2$	$I_3 = \frac{4}{3}$
$R_T = 2.7$	$R_1 = 5$	$R_2 = 10$	$R_3 = 15$
P_T			

For Parallel Circuits

$\boxed{V_1 = V_2 = V_3 = \cdots \text{cte}}$

2nd step $I = ?$

Due to a Parallel Circuit $\boxed{V_T = 20 = V_1 = V_2 = V_3 = 20}$

$\dfrac{1}{R_T} = \dfrac{1}{5} + \dfrac{1}{10} + \dfrac{1}{15} = \dfrac{6+3+2}{30} = \dfrac{11}{30} \Rightarrow R_T = \dfrac{30}{11} = 2.72$

Total Resistance is less than the smallest Resistor

$R_T = 2.7 < 5\Omega$ ✓

$I_1 = \dfrac{V_1}{R_1} = \dfrac{20}{5} = 4A \qquad I_2 = \dfrac{V_2}{R_2} = \dfrac{20}{10} = 2A \qquad I_3 = \dfrac{V_3}{R_3} = \dfrac{20}{15} = \dfrac{4}{3} \qquad I_T = \dfrac{20}{2.72} = 7.35 A$

Mr Lopez

Chapter 5: Types of Circuits

Solving a Parallel Circuit

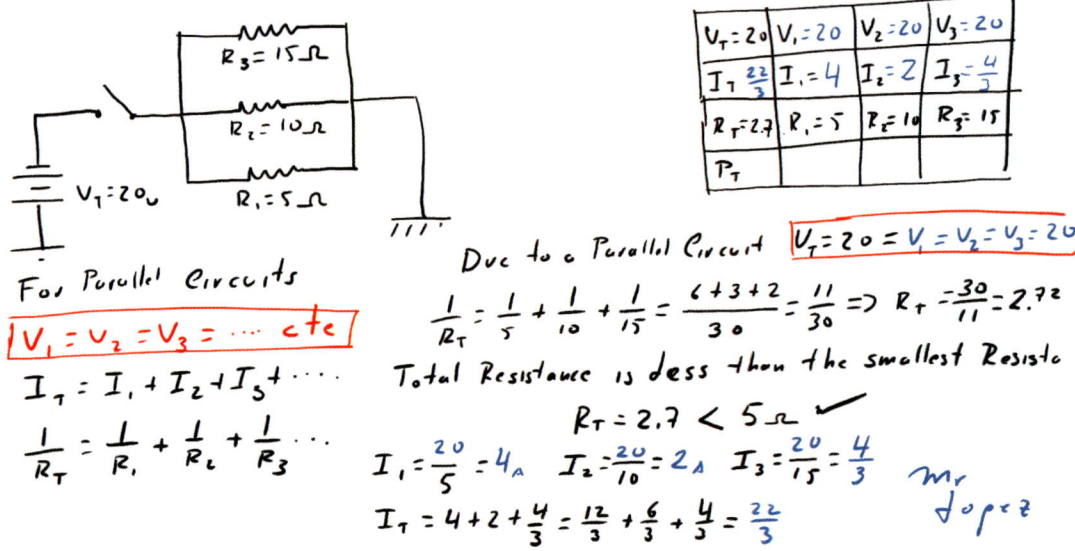

For Parallel Circuits
$V_1 = V_2 = V_3 = \ldots = cte$
$I_T = I_1 + I_2 + I_3 + \ldots$
$\dfrac{1}{R_T} = \dfrac{1}{R_1} + \dfrac{1}{R_2} + \dfrac{1}{R_3} \ldots$

Due to a Parallel Circuit $V_T = 20 = V_1 = V_2 = V_3 = 20$

$\dfrac{1}{R_T} = \dfrac{1}{5} + \dfrac{1}{10} + \dfrac{1}{15} = \dfrac{6+3+2}{30} = \dfrac{11}{30} \Rightarrow R_T = \dfrac{30}{11} = 2.72$

Total Resistance is less than the smallest Resistor

$R_T = 2.7 < 5\,\Omega$ ✓

$I_1 = \dfrac{20}{5} = 4\,A \quad I_2 = \dfrac{20}{10} = 2\,A \quad I_3 = \dfrac{20}{15} = \dfrac{4}{3}$

$I_T = 4 + 2 + \dfrac{4}{3} = \dfrac{12}{3} + \dfrac{6}{3} + \dfrac{4}{3} = \dfrac{22}{3}$

Mr. Lopez

Combined Circuits

Many circuits have a combination of series and parallel resistors. Generally, the total resistance in a circuit like this is found by reducing the different series and parallel combinations step-by-step to end up with a single equivalent resistance for the circuit.

Solving Combined Circuits

The first circuit can be reduced to the second circuit. the secret is to solve the small circuits in parallel and turning them into an equivalent resistance.

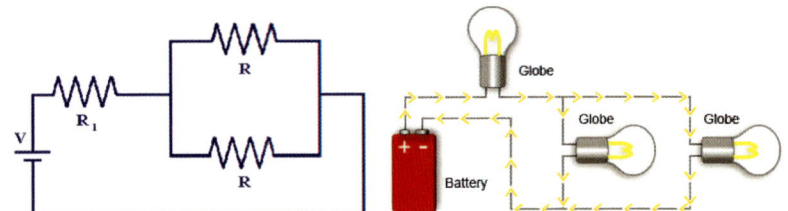

$\dfrac{1}{R_f} = \dfrac{1}{R_2} + \dfrac{1}{R_3} = \dfrac{2+1}{20}$

$\Rightarrow \dfrac{R_f}{1} = \dfrac{20}{3} \Rightarrow R_f = \dfrac{20}{3}$

$R_T = 5\,\Omega + \dfrac{20}{3}\,\Omega = \dfrac{15+20}{3} \Rightarrow$

$R_T = \dfrac{35}{3}\,\Omega$

Series Parallel circuit

To solve the circuit, I recommend follow this procedure

A series-parallel combination circuit

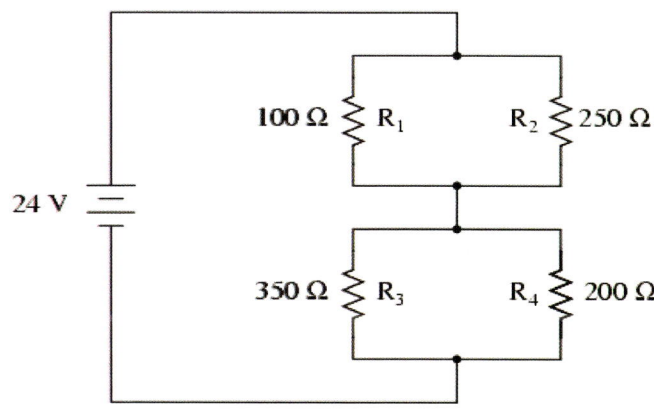

Solving Complex Circuits

Each small parallel circuit can be solved to convert the original circuit into the right circuit.

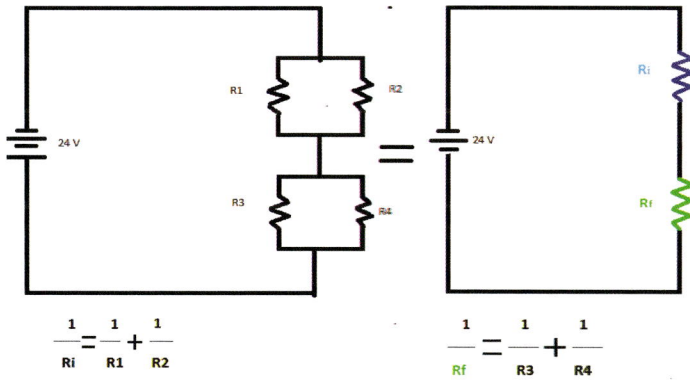

$$\frac{1}{Ri} = \frac{1}{R1} + \frac{1}{R2} \qquad \frac{1}{Rf} = \frac{1}{R3} + \frac{1}{R4}$$

Finally, the original circuit is reduced into a simple series circuit.

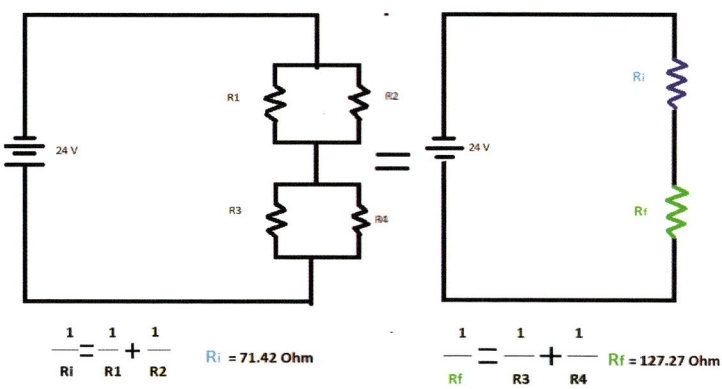

$$\frac{1}{Ri} = \frac{1}{R1} + \frac{1}{R2} \quad Ri = 71.42\ Ohm \qquad \frac{1}{Rf} = \frac{1}{R3} + \frac{1}{R4} \quad Rf = 127.27\ Ohm$$

Chapter 5: Types of Circuits

Solving Complex Circuits

In this example we will solve a combined circuit.
The first circuit can be reduced into the second circuit. the key is to solve the small parallel circuit and turning them into an equivalent series circuit.

Solve the following Series/parallel

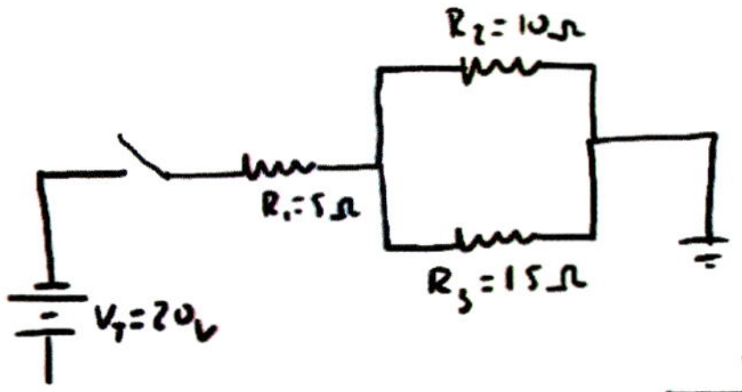

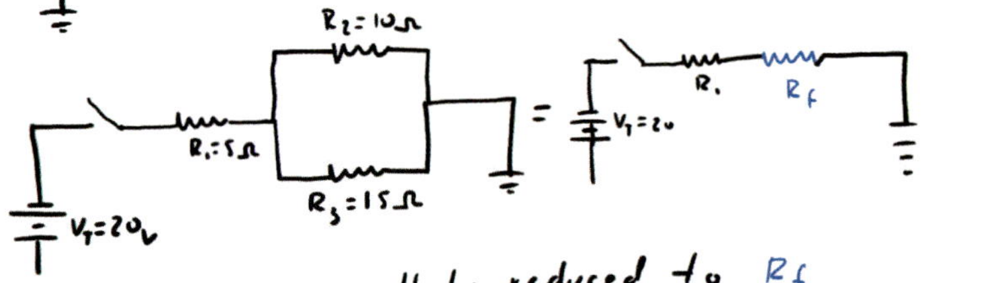

Step 1 ⇒ R_2 y R_3 will be reduced to R_f

$$\frac{1}{R_f} = \frac{1}{10} + \frac{1}{15} = \frac{3+2}{30} \Rightarrow R_f = \frac{30}{5} = 6\,\Omega$$

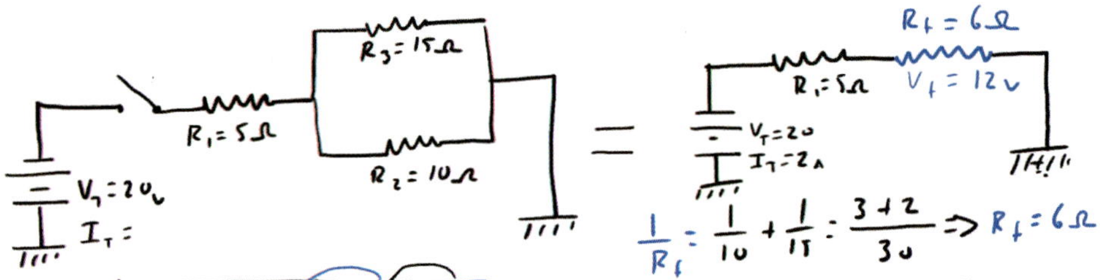

$R_f = 6\,\Omega$

$$\frac{1}{R_f} = \frac{1}{10} + \frac{1}{15} = \frac{3+2}{30} \Rightarrow R_f = 6\,\Omega$$

The Second circuit is a Series ⇒
$R_T = R_1 + R_f = 5\,\Omega + 6\,\Omega = 11\,\Omega$
and $I_T = \frac{20}{11} = 1.81 \Rightarrow = I_1 = I_f = 1.81$

	$V_T=20$	$V_1=$	$V_2=$	$V_3=$	$V_f=$
	$I_T=1.8$	$I_1=1.81$	$I_2=$	$I_3=$	$I_f=1.81$
	$R_T=11$	$R_1=5\Omega$	$R_2=10$	$R_3=15$	$R_f=6$
	P_T				

$$\frac{1}{R_f} = \frac{1}{10} + \frac{1}{15} = \frac{3+2}{30} \Rightarrow R_f = 6\,\Omega$$

The Second circuit is a Series ⇒
$R_T = R_1 + R_f = 5\,\Omega + 6\,\Omega = 11\,\Omega$
and $I_T = \frac{20}{11} = 1.81 \Rightarrow = I_1 = I_f = 1.81$

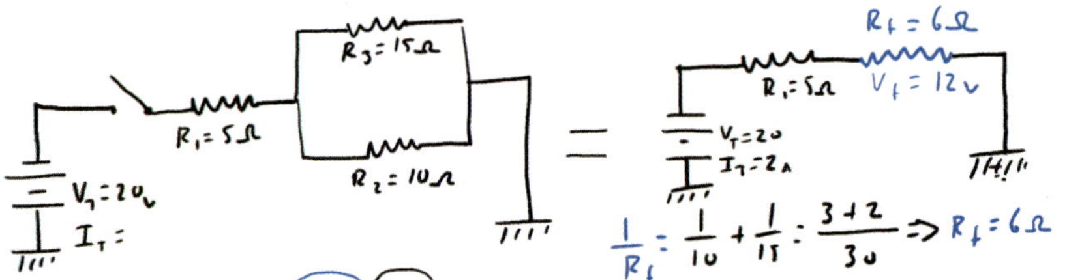

	$V_T=20$	$V_1=9$	$V_2=11$	$V_3=11$	$V_f=11$
	$I_T=1.8$	$I_1=1.8$	$I_2=1.1$	$I_3=0.7$	$I_f=1.8$
	$R_T=11$	$R_1=5\Omega$	$R_2=10$	$R_3=15$	$R_f=6$
	P_T				

⇒ $V_f = I_f \times R_f = 1.81 \times 6 = 11\,v$
⇒ V_f is equal to V_3 and V_2 (Parallel)
$V_f = 11 = V_2 = V_3 = 11\,v$
⇒ $V_T = V_1 + V_f \Rightarrow 20 = V_1 + 11 \Rightarrow V_1 = 9\,v$

Mr. Lopez

Solving Complex Circuits

Because the final circuit is a series circuit the total resistance will be the sum of each resistance.

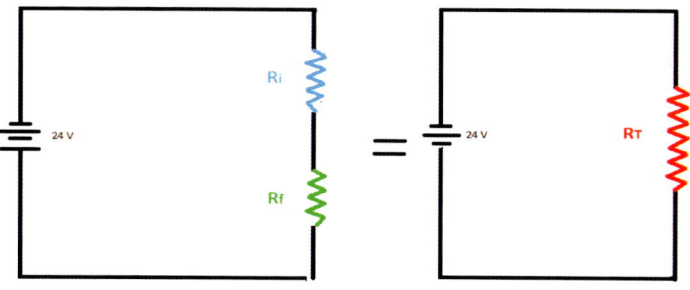

$R_T = R_i + R_f = 71.42 + 127.27 = 198.69$ Ohm

Tips to Solve Complex Circuits

To solve the rest of the circuit I recommend completing the following table step by step

VT = 24V	V1 = ?	V2 = ?	V3 = ?	V4 = ?
IT = ?	I1 = ?	I2 = ?	I3 = ?	I4 = ?
RT = **198.69**	R1 = 100	R2 = 250	R3 = 350	R4 = 200
PT = ?	P1 = ?	P2 = ?	P3 = ?	P4 = ?

Remember that series circuits have a constant element, **the intensity** (I=Amp)
Using Ohm Law IT = VT / RT = 24 / 198.7
IT = **0.12 Amp**
IT = Ii = If = **0.12 A** (Series Circuit)

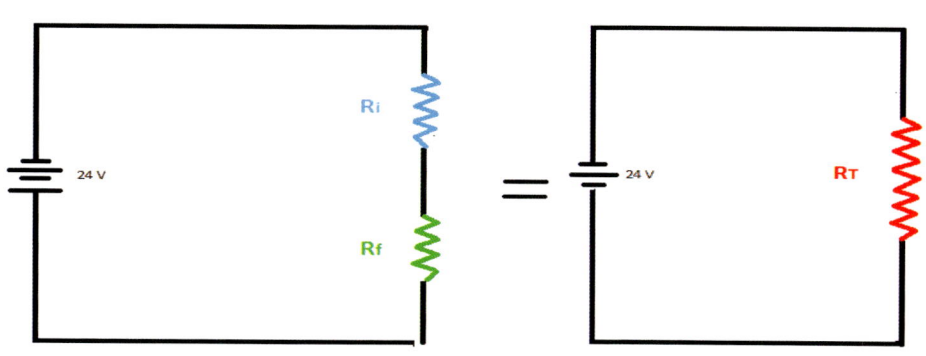

$R_T = R_i + R_f = 71.42 + 127.27 = 198.69$ Ohm

Tips to Solve Complex Circuits

IT = Ii = If = **0.12 A** (Series Circuit)
Vi = Ii X Ri = 0.12 X 71.42 = **8.6 Volts**
Vf = If X Rf = 0.12 X 127.27 = **15.3 Volts**

Now remember that parallel circuits have a constant element, **the voltage.**

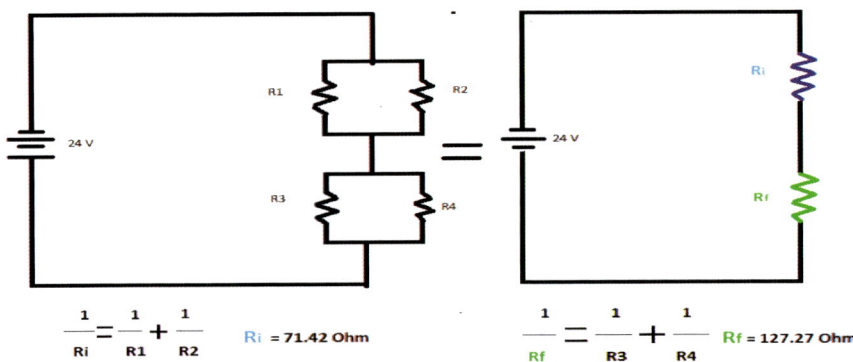

Vi = V1 = V2   and   **Vf = V3 = V4**
V1 = V2 = 8.6 Volt   and   V3 = V4 = 15.3 Volt
Again using Ohm's law we can calculate the intensities at points 1, 2, 3 and 4
 – I1 = V1 / R1 = 8.6/100 = 0.086 Ohm
 – I2 = V2 / R2 = 8.6/250 = 0.034 Ohm
 – I3 = V3 / R3 = 15.3/350 = 0.043 Ohm
 – I4 = V4 / R4 = 15.3/200 = 0.076 Ohm

To calculate the power we follow the same procedure
PT = VT x IT = 24 x 0.12 = 2.88 W

Final Result

To solve the rest of the circuit I recommend completing the following table step by step

VT = **24V**	V1 = 8.6V	V2 = 8.6V	V3 = 15.3VV	V4 = 15.3V
IT = 0.12A	I1 = 0.086A	I2 = 0.034A	I3 = 0.043A	I4 = 0.076A
RT = **198.69**	R1 = **100**	R2 = **250**	R3 = **350**	R4 = **200**
PT = 2.88W	P1 = 0.73W	P2 = 0.29W	P3 = 0.65W	P4 = 1.16W

Power

According to Ohm's law, power is defined as the product of the voltage by the current:
P(watts) = V(Volt) X I(Amp)

The unit of measure is in [Watts]   **1 KW = 1000 Watts**

12 V or 24 V ?

Some boat owners argue that their boat has more Power because it uses 24 volts compared to other boats that use 12 Volts.
Is this true?

To answer this question we will analyze what happens when we connect batteries in series or parallel.

Series Connection (24V)

$$V_t = V_1 + V_2 = 24v$$
$$I_T = I_1 = I_2 = 100 \, Amp$$
$$R_T = R_1 + R_2 \; ohm.$$

$$P_T = V_T \times I_1 = 24v \times 100 Am = \underline{2400 \, watts}$$

Chapter 5: Types of Circuits

Parallel Connection (12V)

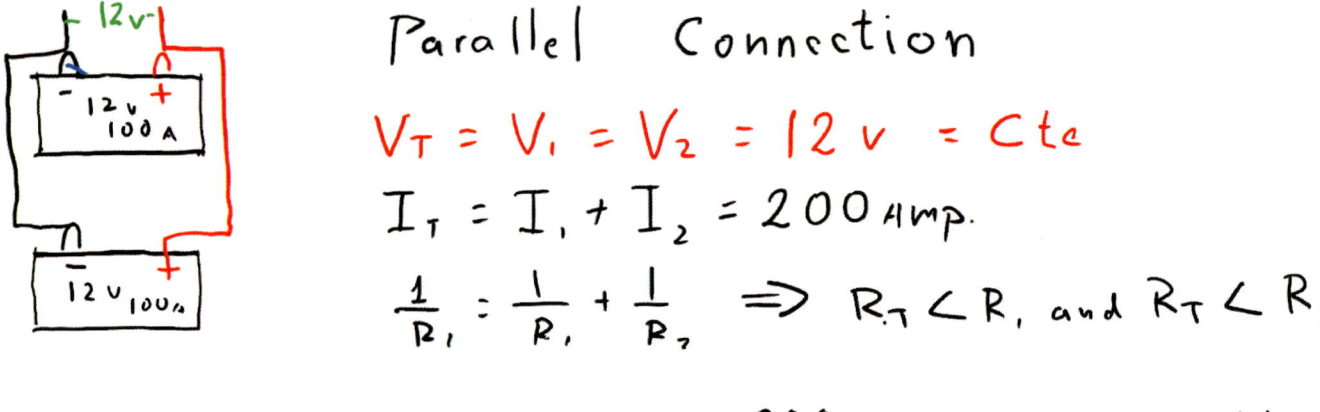

Parallel Connection

$$V_T = V_1 = V_2 = 12v = Cte$$
$$I_T = I_1 + I_2 = 200 \, Amp.$$
$$\frac{1}{R_T} = \frac{1}{R_1} + \frac{1}{R_2} \Rightarrow R_T < R_1 \text{ and } R_T < R_2$$

$$P_T = V_T \times I_T = 12v \times 200 \, Am = \underline{2400 \, watts}$$

Parallel connections are recommended where you need storage the maximum Amp per Hour for example in:

- House Battery Banks
- Bow and Stern Thruster Banks
- Solar cells arrays

Of course those parallel connections could be connected in series with other group of batteries the get the Voltage and Ampacity required.

In general Deep Cycle batteries are recommended in Parallel configurations where you need deliver the charge slowly.

Power (12 Volt or 24 Volt ?)

As you can see the power is the same 2400 Watts in both cases
However, the intensity is greater when we make parallel configurations
IT = 200A (Parallel)   IT = 100A (Series)
For this reason, the conductor size must be greater for parallel configurations. (See Chapter 4 - Wire Size Calculation)

Chapter 5: Types of Circuits

Power (12 Volt or 24 Volt ?)

Parallel 12v

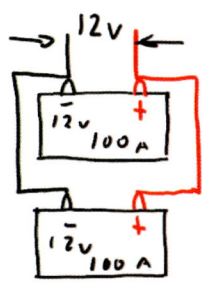

$V_T = 12v =$ Constant
$I_T = 200$ amp
$P_T = V_T \times I_T = 12 \times 200 = \underline{2400 \text{ Watts}}$

Series 24v

$V_T = 12v + 12v = 24v$
$I_T = 100$ Amp $=$ Constant
$P_T = V_T \times I_T = 24 \times 100 = \underline{2400 \text{ W}}$

The Power is the Same in both Configurations 12v or 24v

Analysis:
In 12v configuration there are 200 amps. However
In 24v configuration there are only 100 amps
Then: In 12v we need **thicker wires** due to the flow of electrons per hour 200 Amp/hr Vs 100 Amp/hr in Series Conf.

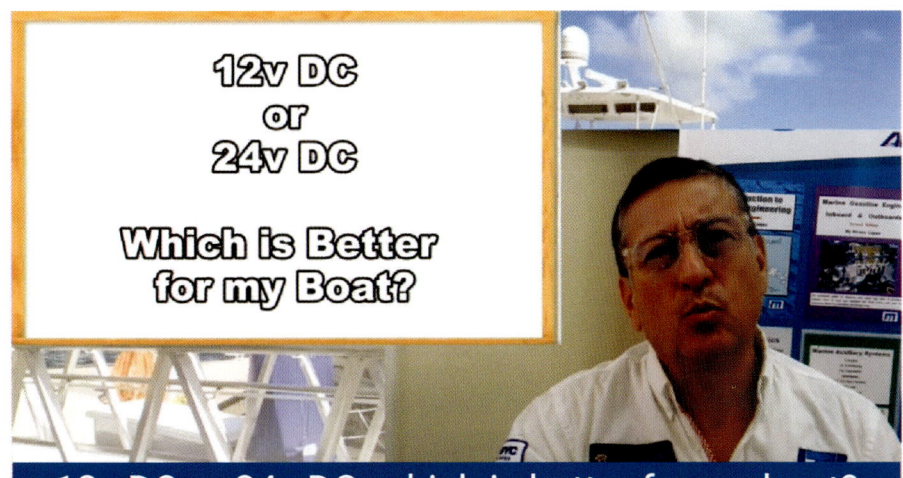

12v DC or 24v DC, which is better for my boat?

CHAPTER 6

Charging & Cranking Systems

TOPICS

6.0 Types of Switches	75
6.1 Push Bottom Switch	76
6.2 Relays	76
6.3 Solenoids	77
6.4 Battery Switch Selector	79
6.5 Battery Isolator or ACR	81
6.6 The Alternator	83
6.7 Marine Battery Charger	87
6.8 Selecting the Charger	88
6.9 Cranking Systems	89
6.10 The Starter Motor	89
6.11 Starter Troubleshooting	91

Video Episode 1: Relays and Solenoids

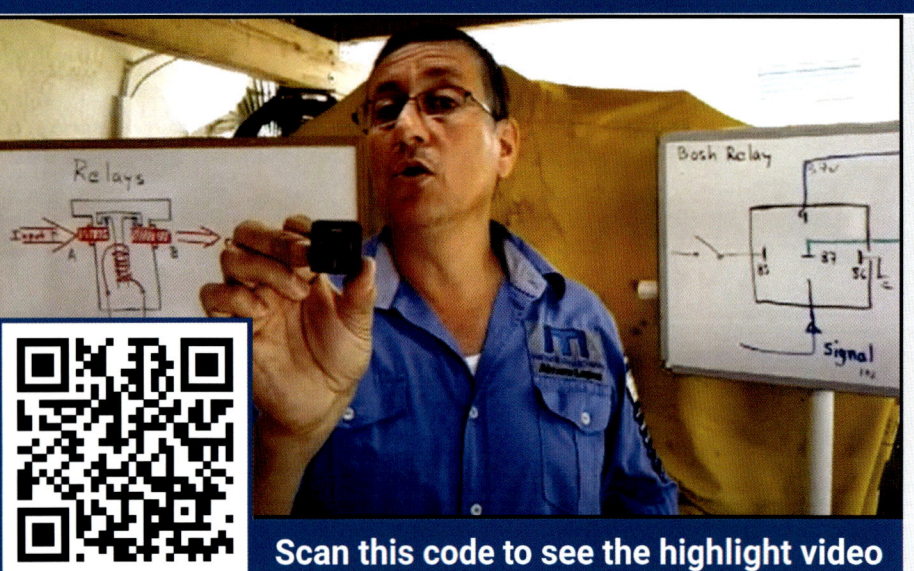

Scan this code to see the highlight video

In this video you will learn the procedure to diagnosis relays and solenoids in a simple way. Also we are going to study different applications for relays and solenoids.

Follow me

Switches

Mechanical switches permit or interrupt the flow of current. This is called an SPST (Single-Pole, Single-Throw)

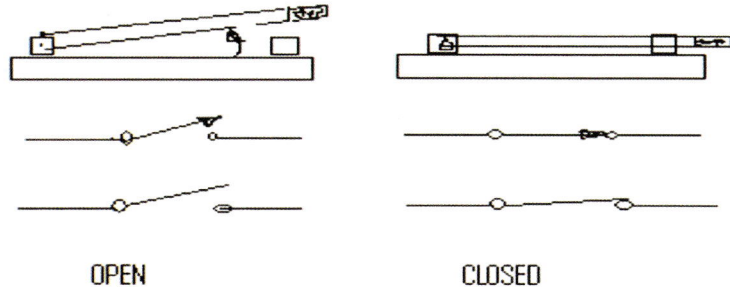

Push Button Switch

The normally open (NO) switch is used to send the signal to the starter solenoid.
The normally close (NC) switch is used to interrupt the signal that feeds the fuel system.

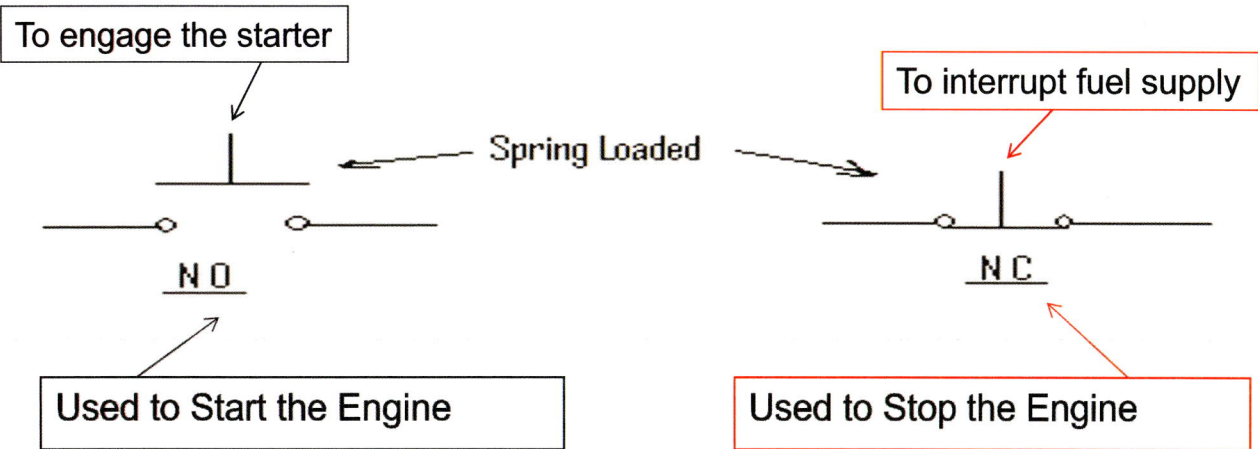

Usually SPST, Normally Open (NO) or Normally Closed (NC)

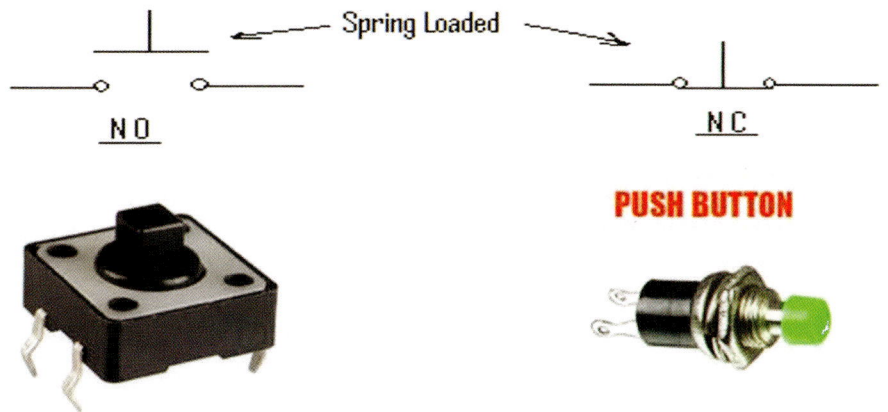

Multiple Contact Switches

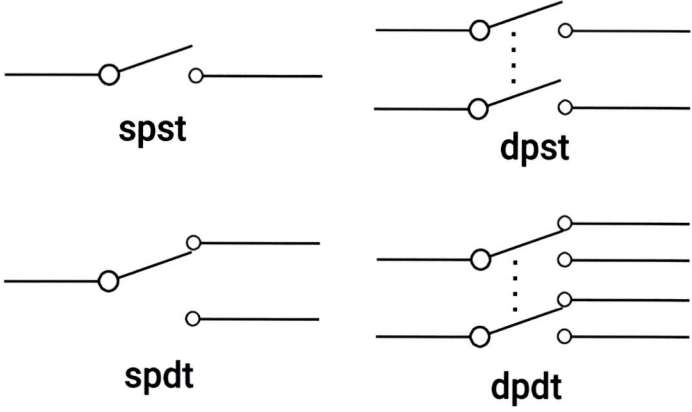

SPDT : Single-Pole, Double-Throw
DPST : Double-Pole, Single-Throw
DPDT : Double-Pole, Double-Throw

Push Button Switch

Some diesel engines use a momentary button to open or close the circuits that start or stop the engine.
The momentary buttons are, Normally Open (NO) or Normally Closed (NC).

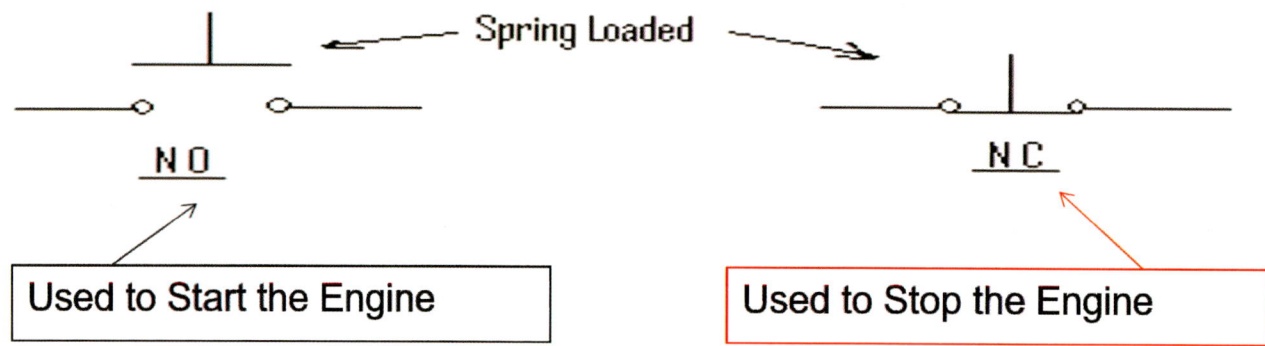

Relays

A **relay** is an electrically operated switch.
Relays are used where it is necessary to control a circuit by a low-power signal (with complete electrical isolation between control and controlled circuits).

High Current Relay

With a small signal coming from the ignition switch the relay allow a high current flow between A and B terminals.

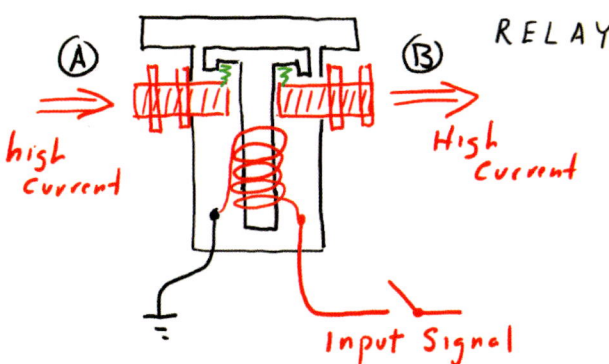

Relay for Starter Circuit

When the ignition key is turned all the way to the "start" position, it allows electricity to flow to the starter solenoid (relay) which then connects the battery to the starter motor.

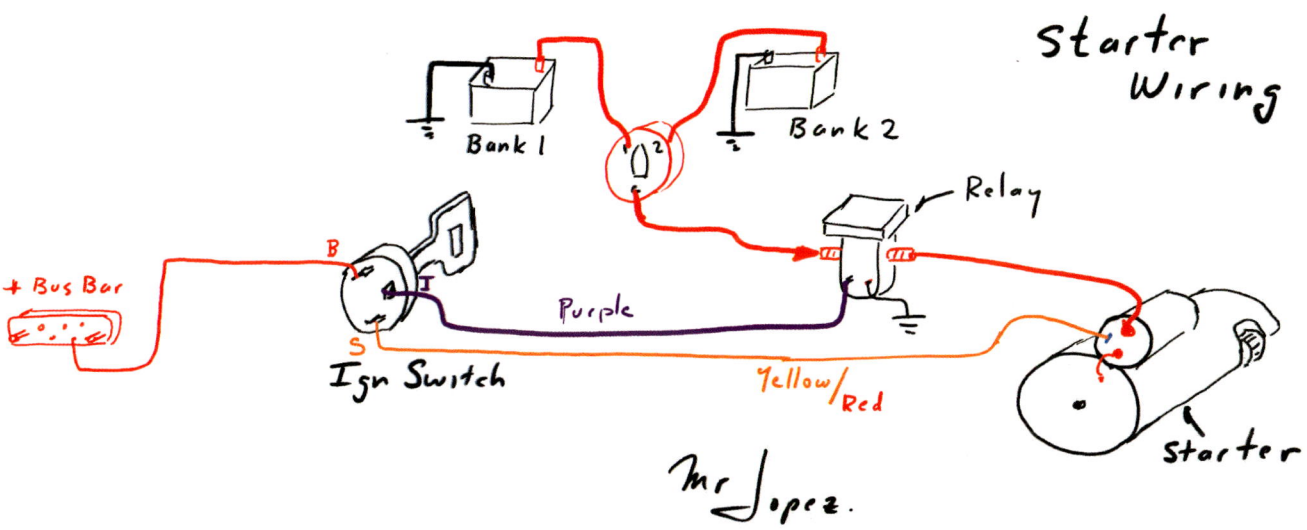

Solenoids

Solenoids are actuators capable of linear motion. They can be electromechanical (AC/DC), hydraulic, or pneumatic driven - all operating on the same basic principles.
They are great for pushing buttons, hitting keys on a piano, valve operators, and even for jumping robots.
The difference between a solenoid and a motor is that a solenoid is spring loaded and cannot rotate.

How Solenoids Work?

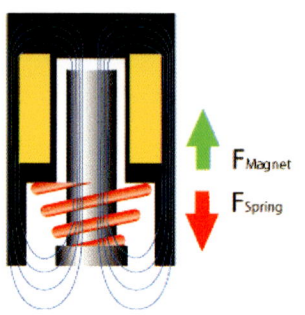

The inner shaft of a solenoid is a piston like cylinder made of iron or steel, called the **plunger** or **slug** (equivalent to an **armature**).

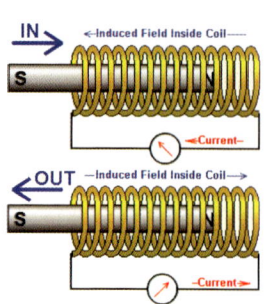

Magnetic Field

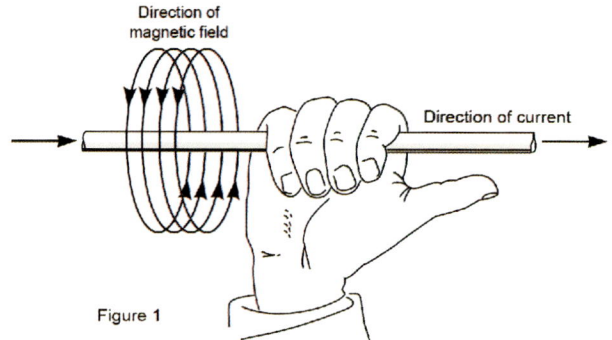

Figure 1

When a current pass through a conductor, produce a magnetic field around the wire following the right-hand rule (Figure 1)

Relay / Solenoid

Both work on the same principle of operation the only difference is that the solenoid produces an external mechanical movement.

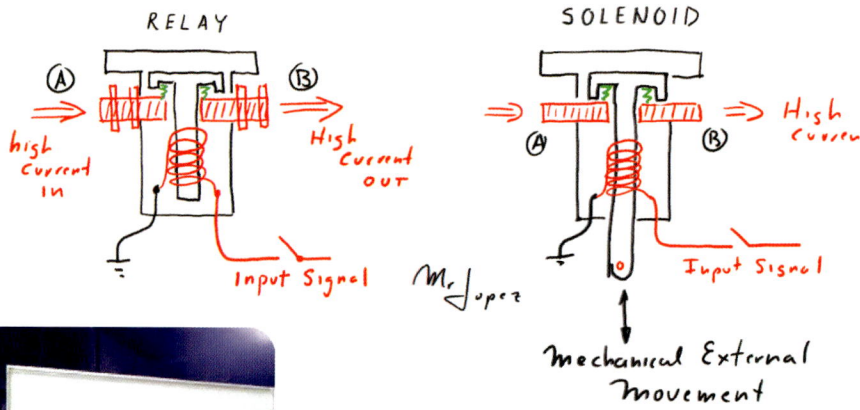

Relays and Solenoids

Chapter 6: Charging & Cranking Systems

Video Episode 2: Battery Switch Selector and ACR

In this video you will learn the fundamentals about the Battery Switch selector and the (A.C.R) Automatic Charging Relay. and how those elements work together in order to keep the charge level equal for two battery banks.

Follow me

Scan this code to see the highlight video

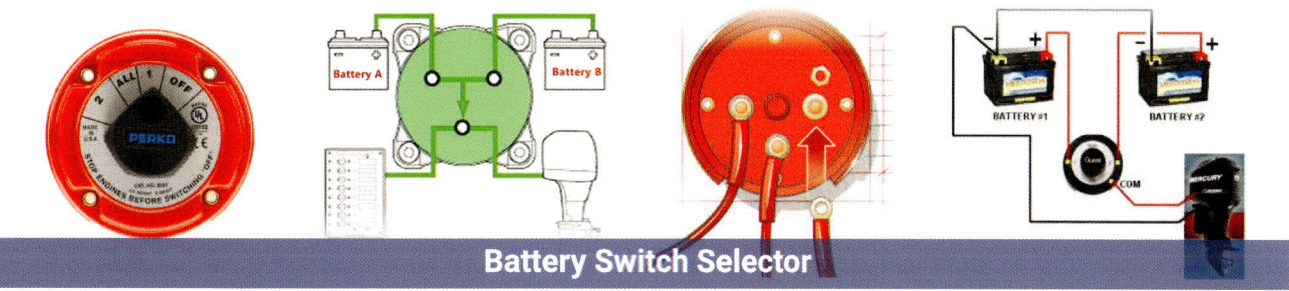

Battery Switch Selector

Battery Switch selector Types

Basically, there are two different ways (with a few variations) folks usually wire them.
One way is with each battery separately being able to handle all the boat's needs and you just switch between them.

The other has two different type of batteries, one used for starting purposes (**Crank Battery**) and one used for running the boat's system (**House Battery**).

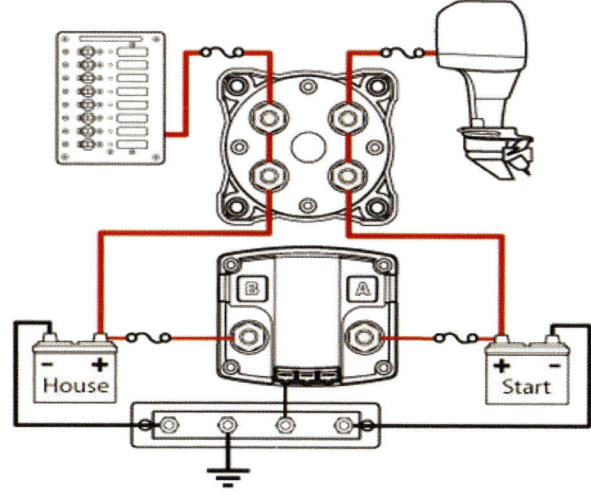

Battery Switch selector Types

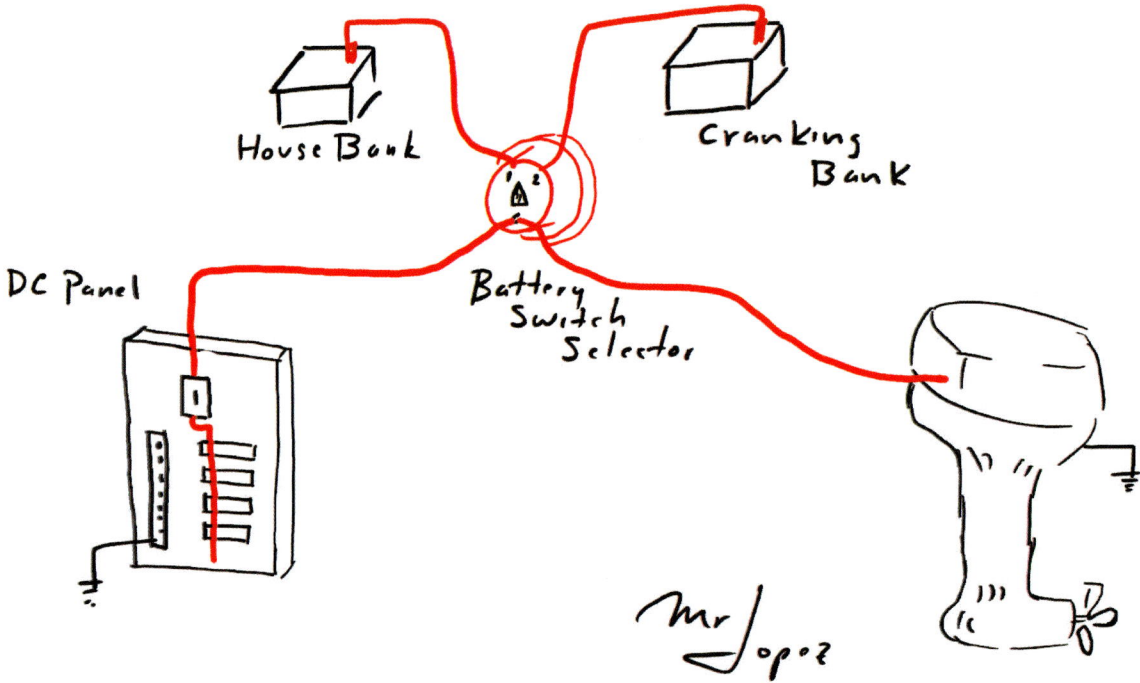

Battery Switch

Normally the people use one battery one time the other the next time. This way the other battery is a true back up.

Single engine –Two banks

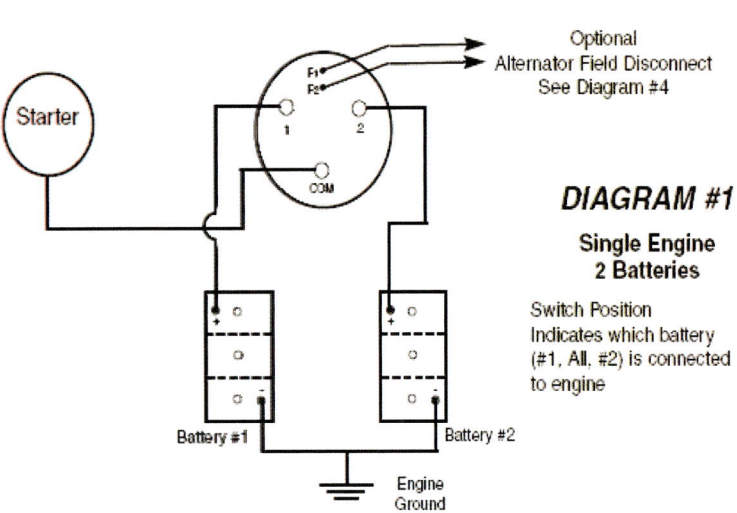

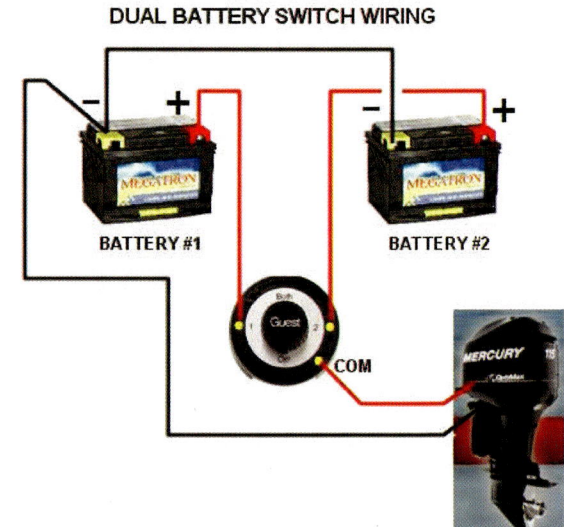

The Battery Isolator (1 Engine)

Its function is to keep the load on both batteries equally.

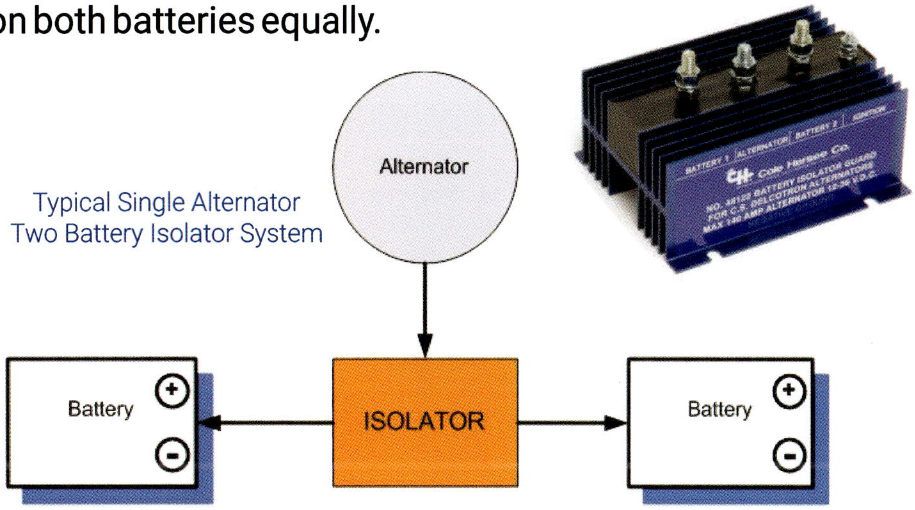

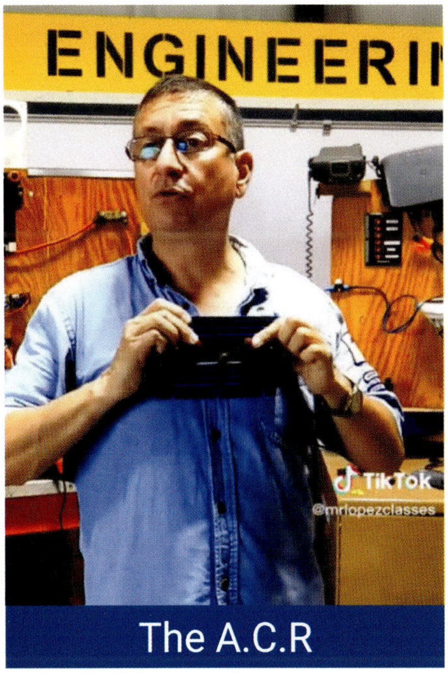

Typical Single Alternator Two Battery Isolator System

Battery Stabilizer (2 Positions)

One outboard engine and House panel using separate batteries. In this example the engine alternator charge both battery banks through the ACR.

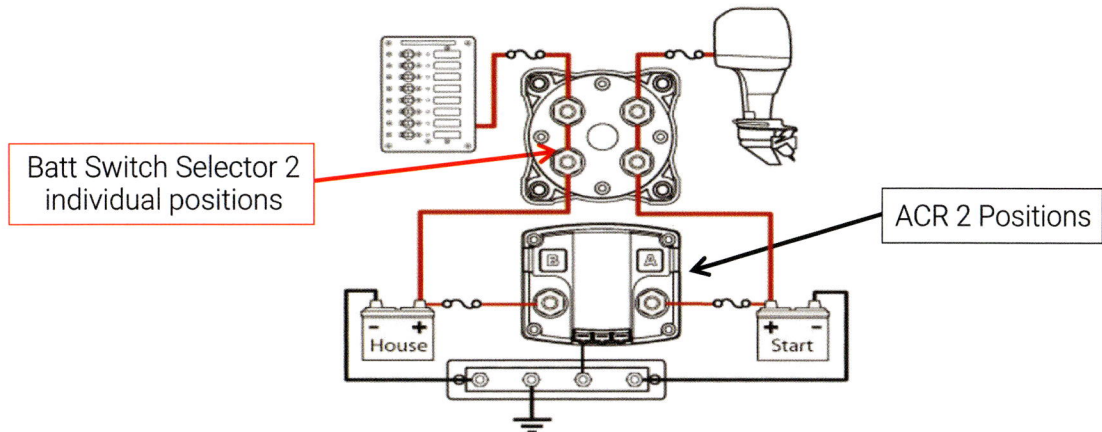

Chapter 6: Charging & Cranking Systems

Battery Isolator or A.C.R (2 Inputs)

Two Inboard engines that can be started by either battery bank. If one alternator is broken the other one keep both battery banks fully charged through the ACR.

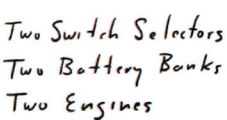
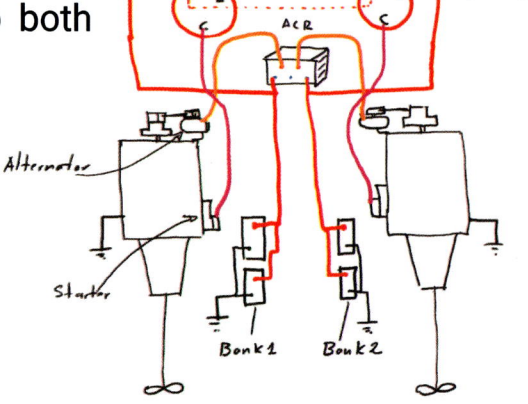

Battery Isolator or A.C.R (1 Input 2 Outputs)

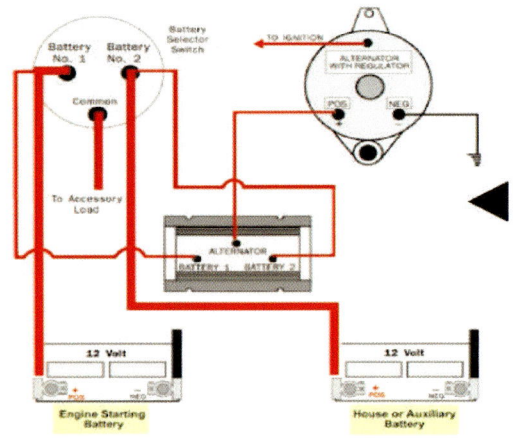

In this case the alternator of a single inboard engine is used to charge both batteries.

Battery Isolator or A.C.R (3 Inputs 4 Outputs)

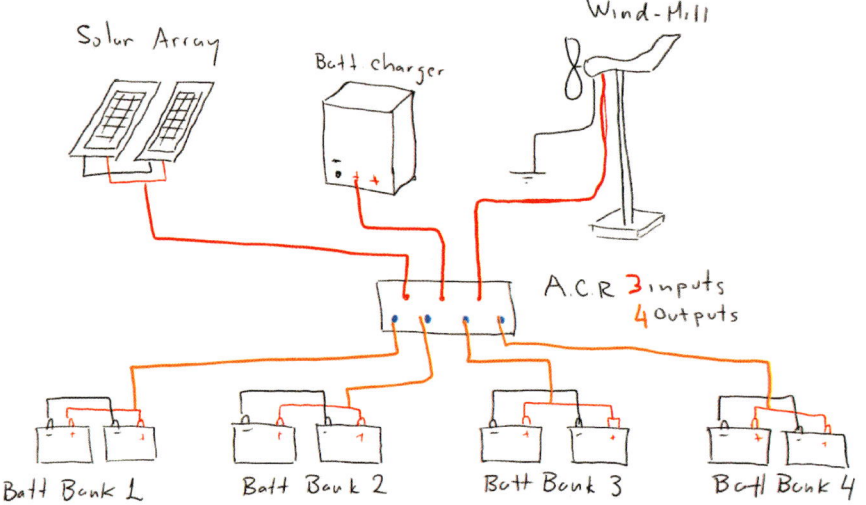

Chapter 6: Charging & Cranking Systems

Video Episode 3: Marine Alternators Fundamentals and Troubleshooting

In this video you will learn the fundamentals about Marine Alternator and how the Alternator, The Solar Panels and the Battery chargers are integrated such as sources of DC current. The video we will learn the diagnosis about the typical fault codes in marine alternators.

Scan this code to see the highlight video

Follow me

Marine Alternators Fundamentals and Troubleshooting

The Alternator

An Alternator is an electro-mechanical device that converts mechanical energy into electrical energy. In pleasure yachts, the alternator is vital to keep different battery banks charged.

Main Components

The majority of the alternators use a rotating magnetic field with a stationary armature. The principle of operation is magnetism.

Chapter 6: Charging & Cranking Systems

Magnetism

Suppose you have a steel bar. Now let a coil of wire around the bar. Due to the magnetic properties the iron, the positive charges and the negative charges will be organized at the ends of the bar.

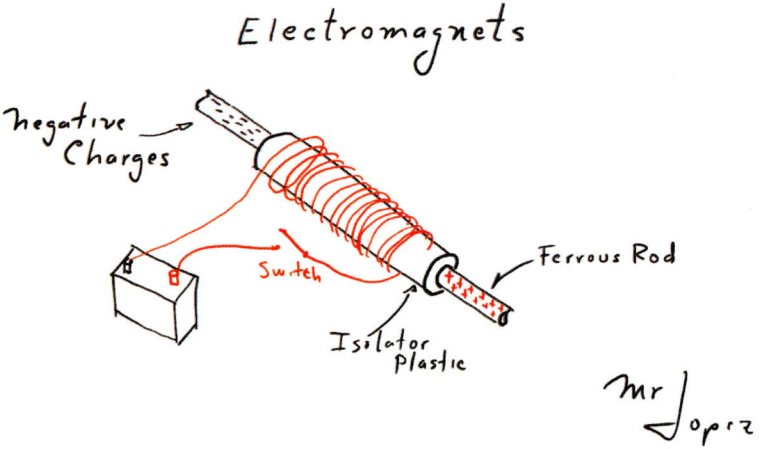

Principle of Operation

Now we will weld a disc the same material on each side of the bar and then we'll cut small bites around each disk.

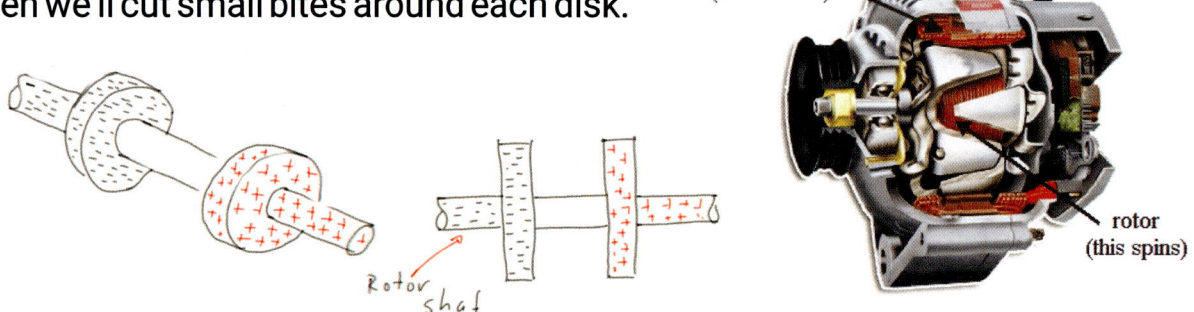

Then we will wrap copper wire to create a coil in the center of the bar. Finally, we fold the pre-cut sections of each disk so that the coil falls within the two discs.

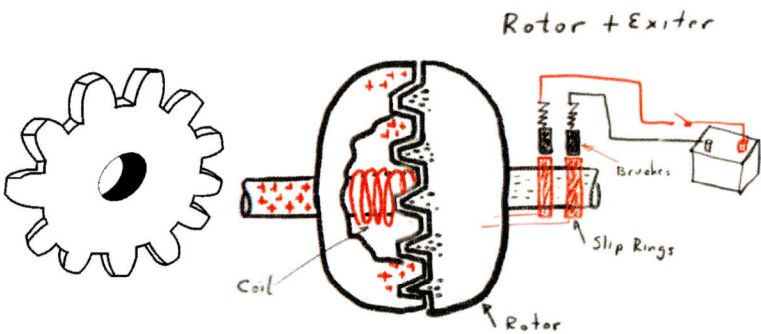

The Rotor Magnetic Field

The coil wire is connected with the slip rings and these receives the signal from the ignition switch through the brushes.

AC Signal

When the rotor is turning the positive and negative charges are thrown by centrifugal force and are received by the stator winding, producing a sinus signal AC.

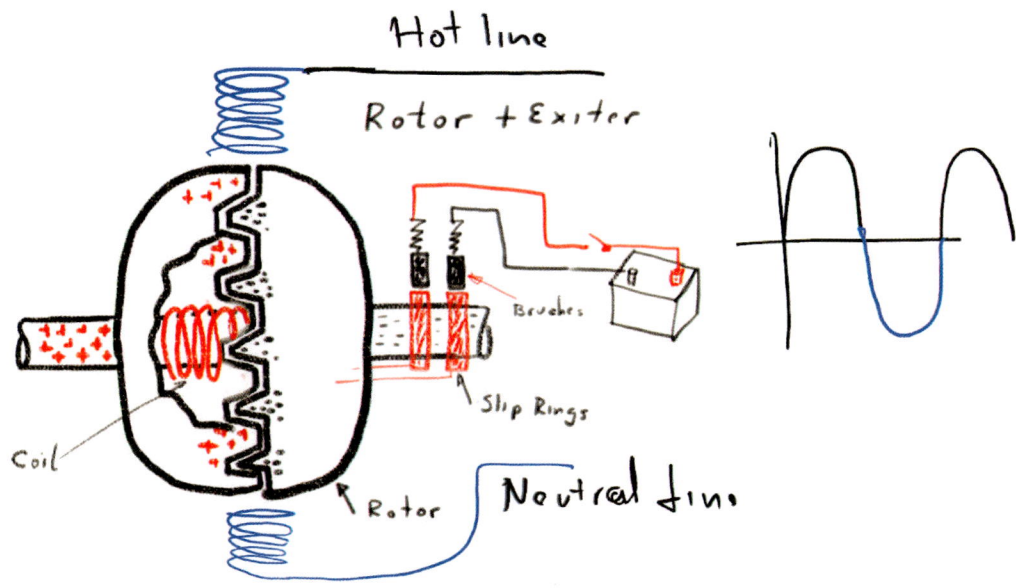

AC Output Signal

Now you can read an AC signal in the three output terminals of the stator. At this point we have an AC generator.

Hot, Neutral and Ground terminals

AC to DC Rectification

The next step involves passing the alternating current through a device called *the bridge of diodes* that converts it into direct current.

Voltage Regulator

An automatic voltage control device controls the field current to keep output voltage constant. If the output voltage from the stationary armature coils drops due to an increase in demand, more current is fed into the rotating field coils through the voltage regulator (VR)

Chapter 6: Charging & Cranking Systems

Marine Battery Chargers

Video Episode 4: Marine Battery Chargers

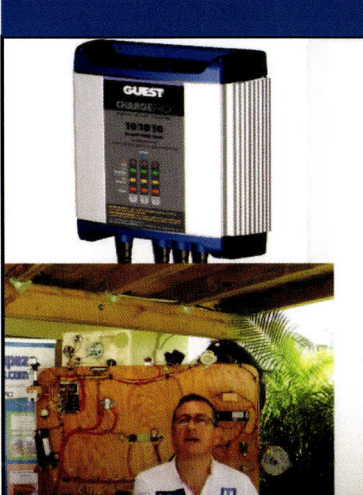

Scan this code to see the highlight video

Selecting the Charger

- The charger should ideally offer battery temperature sensing and come with the sensor as standard equipment, not an "extra".
- The charger should include multiple options for charging voltages/programs
- The charger needs to h[ave] and the manufacturer [has a] reputation for customer [service]

In this video you will learn the procedure to select the appropriate Battery Charger according with the capacity of the battery bank in Amp/Hr and the procedure to size the charger and alternator appropriate per each battery bank.

Follow me

In simple words, a battery charger is a sophisticated rectifier. This rectifier converts Alternate Current to Direct Current.
The charger should work on varying input voltages and not suffer from output limiting.

Input 120/240 VAC   Output 12/24 VDC

Select the proper battery charger according to the capacity of the battery bank in Amp per hour, the type of battery (Lead Acid, Gel, AGM, Lithium, etc), and the Voltage

The general consensus is to select a battery charger with a capacity in Amp of 10% of the Battery Bank capacity bank's capacity. In this case, a 300Ah battery bank would work properly with a 30A charger.

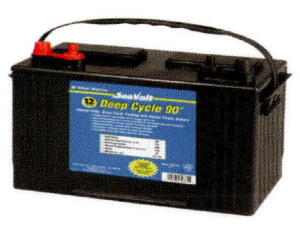 

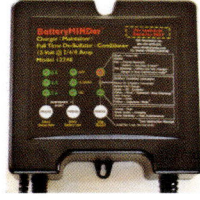 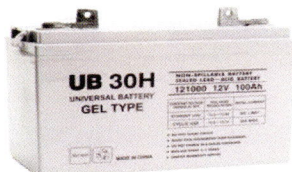

Use the correct charger according to the battery type (GEL, AGM or Lead Acid). improper charger can ruin the batteries immediately.

Periodically it is advisable to remove the output terminals of the charger and verify that the voltage produced is above 14 V.

87

Chapter 6: Charging & Cranking Systems

Inverters and Chargers Fundamentals

Selecting the Charger

The charger should ideally offer battery temperature sensing and come with the sensor as standard equipment, not an "extra". The charger should include multiple options for charging voltages/programs.
The charger needs to have a good warranty and the manufacturer should have a good reputation for customer service/support.

Battery Charger and A.C.R

I recommend connecting the battery charger outputs at the input of an ACR to avoid battery overcharging and also to protect the charger.

Warning !!!

For no reason try to install an automotive battery charger on your boat. Car chargers when used on boats can be one of the worst offenders & cause of stray current corrosion due to their internal architecture which very often does not isolate AC & DC sides like a UL Marine charger will.

Cranking Systems

Video Episode 5: Cranking Systems

In this video you will learn about the cranking systems for inboard and outboard gasoline and diesel engines. The last part of the video is related with the diagnosis process.

Follow me

Scan this code to see the highlight video

The Starter Motor

The starter motor converts electrical energy to mechanical energy and is mounted on the cylinder block in a position to engage a ring gear on the engine flywheel.

Starter Common Faults

Most of the faults presented at starters are due to:

- Inappropriate selection of batteries.
- Loose connections
- Corrosion of conductor stranding
- Poor quality at the connections
- Undersized conductors
- Bad reference ground
- Excessive amperage due to hydraulics of the engine.

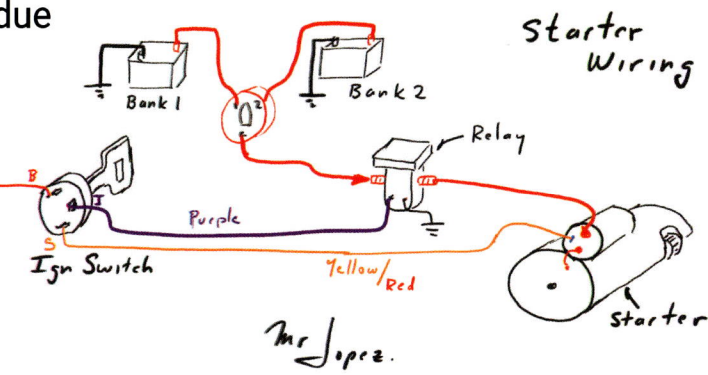

Chapter 6: Charging & Cranking Systems

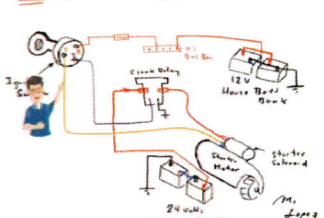

Ignition switch 12 volts and starter 24 volts

Air Plane Engine Starter

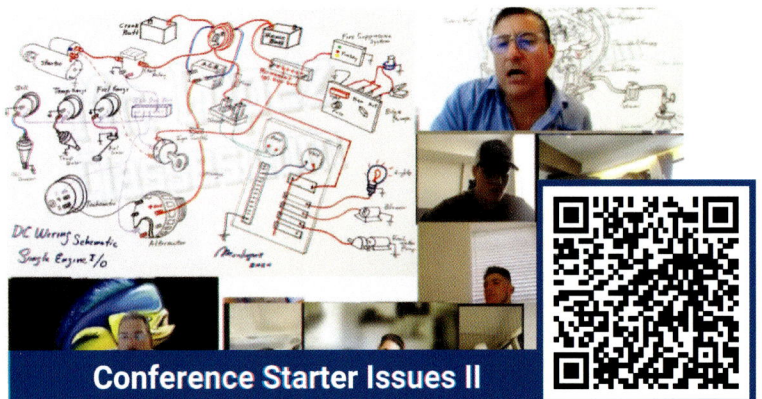

Conference Starter Issues II

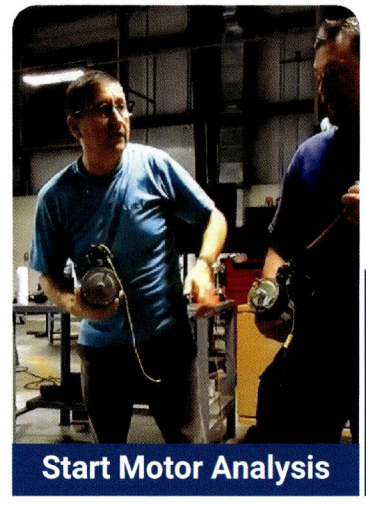

Start Motor Analysis

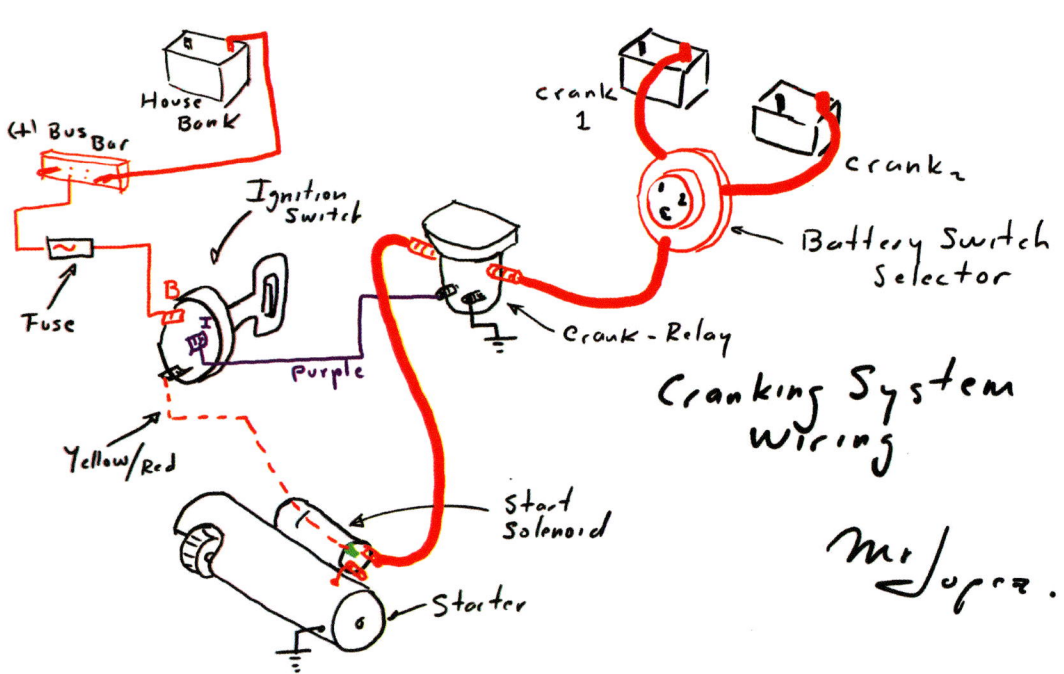

Cranking System Wiring

Troubleshooting

Power (12 /24 vdc) constantly is available from the DC Panel to the ignition switch, as well as to the starter solenoid.

When the ignition switch is turned to the start position, power is available for the ignition circuits, including the ignition and injection computers and to the starter relay in the power panel.

The starter relay passes the power to the starter solenoid. The starter solenoid closes a set of heavy switch contacts, connecting the starter motor to the heavy battery cable. The starter should spin the engine.

Starter Doesn't Spin

If the starter doesn't spin, we can split the system by assuming either:

1. Sufficient power isn't getting to the starter solenoid
2. Power is getting to the solenoid, but sufficient power is not getting to the starter motor.

Once verified that the current is entering the solenoid. Verify that the signal from the ignition switch is coming. The simple click of the solenoid is an indication that the signal is coming from the switch.
Check that the ground is properly bolted at the engine block as close as possible to the starter.

Starter Troubleshoot

In some cases all the parameters are right however you only hear a click but the engine is not spinning.
In this case I recommend try to turn the crankshaft of the engine with a socket and a ratch directly on the harmonic balancer.
If the engine is not turning it is an indication that the engine is locked due to the ingestion of salt water in the combustion chamber.

If the engine is a Gasoline engine; remove the spark plugs and check which cylinder has water
If the engine is Diesel; remove the injectors to verify in what cylinder the water locked the engine.

Conference Cranking Issues

CHAPTER 7

DC Panel

TOPICS

7.1 DC Circuit Breakers	93
7.2 The Ammeter	95
7.3 The Shunt Ammeter	96
7.4 Reading Amperes	97
7.5 The Voltmeter	98

Scan this code to see the highlight video

Video Episode 1
Wiring a boat battery bank

In this video you will learn the procedure step by step to do the DC wiring in a pleasure yacht. Starting on the battery banks configuration, then the installation of battery switch selectors and A.C.R and finally the selection and configuration of the main DC Panel.

Follow me

DC Panel

The DC Panel normally is composed by:

* Main Breaker 　　* Individual Breakers

* Ammeter 　　* Voltmeter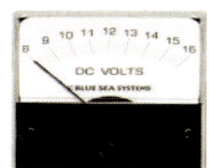

* Positive Bus Bar　　* Negative Bus Bar

* Shunt Ammeter 　　* Battery Switch selector

* Bank Load Monitor

* Main Fuse

DC Circuit Breaker

A DC **circuit breaker** is a manually or automatically operated electrical switch designed to protect an electrical circuit from damage caused by **overload** or **short circuit.**

Unlike a fuse, which operates once and then must be replaced, a circuit breaker can be reset (either manually or automatically) to resume normal operation.

The circuit breaker must detect a fault condition; in low voltage circuit breakers this is usually done within the breaker enclosure.

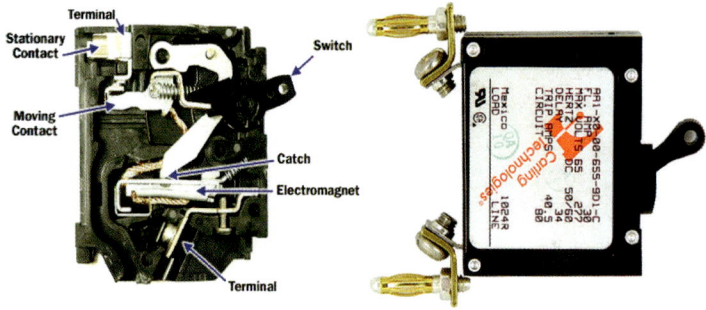

The capacity of the main breaker should be higher or equal to the total capacity of the system (10% higher).

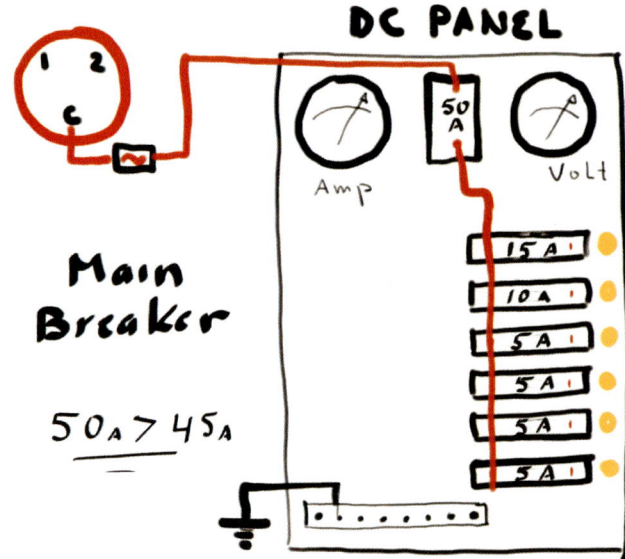

Main Breaker

50ₐ > 45ₐ

Chapter 7: DC Panel

The Ammeter

An **ammeter** is a measuring instrument used to measure the current in a circuit.
Electric currents are measured in amperes (A), hence the name. Instruments used to measure smaller currents, in the milliampere or micro-ampere range, are designated as *milliammeters* or *microammeters*.

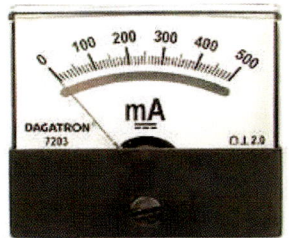
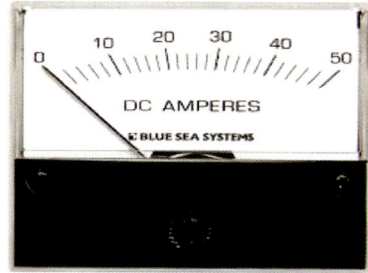

Ammeter Connection

Current [A] is the measure of the rate of electron "flow" in a circuit. The most common way to measure current in a circuit is to break the circuit open and insert an "ammeter" in series (in-line) with the circuit so that all electrons flowing through the circuit also have to go through the meter.

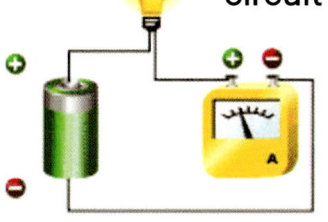

Some digital meters, like the unit shown in the illustration, have a separate jack to insert the red test lead plug when measuring current.

Reading Amps

Warning !!!

An ammeter will act as a short circuit if placed in parallel (across the terminals of) a substantial source of voltage. If this is done, a surge in current will result, potentially damaging the meter.
If the ammeter is accidently connected across a substantial voltage source, the resultant surge in current will "blow" the fuse and render the meter incapable of measuring current until the fuse is replaced. **Be very careful to avoid this scenario!**

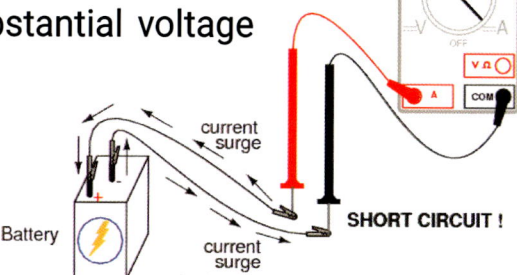

Example

Building the same circuit on a terminal strip should also yield similar results.

The Shunt Ammeter

An **ammeter shunt** is a very low-resistance connection between two points in an electric circuit that forms an alternative path for a portion of the current.

Shunt voltage drop is used in conjunction with an ammeter to measure amperage of a circuit.

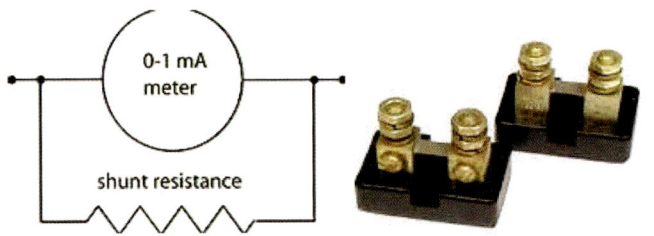

Should be connected between the positive bus bar and the battery switch as close as possible to the ammeter gauge. there are two thin wires coming out of the base and go to the ammeter terminals.

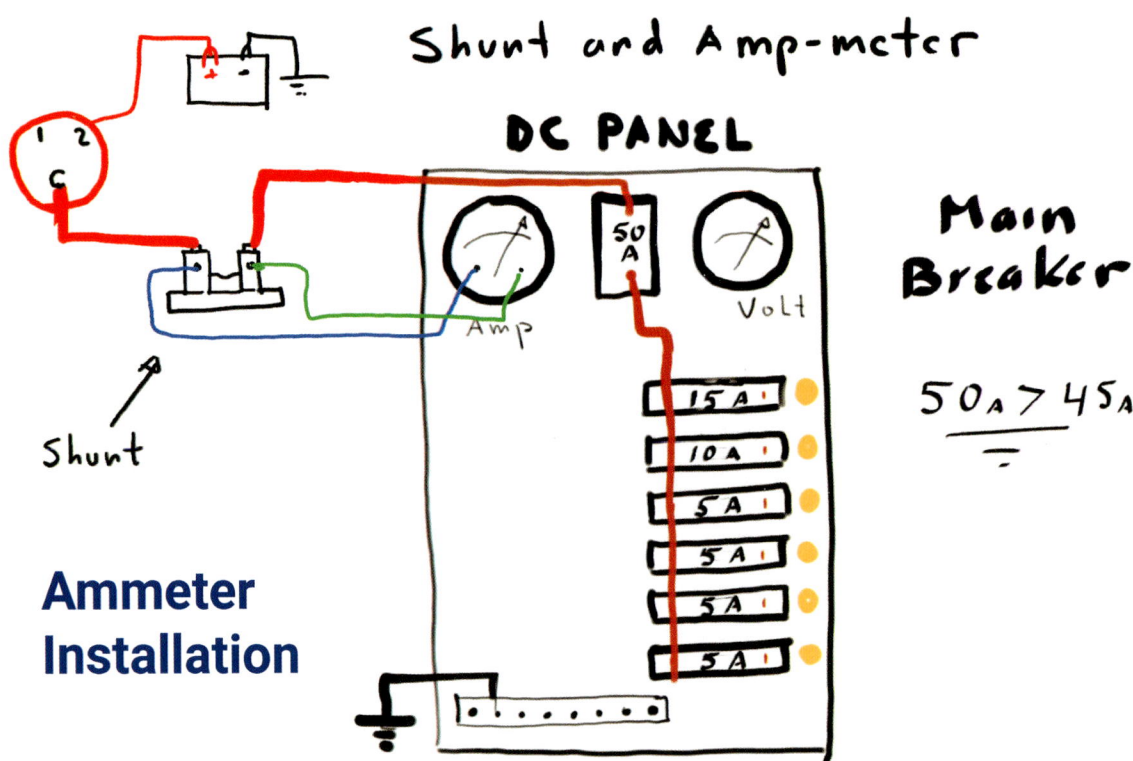

Ammeter Installation

Reading Amps

Break the circuit open at any point (at the breaker , whit the contacts open for example) and connect the meter's test probes to the two points of the break to measure current if your meter is manually-ranged, begin by selecting the highest range for current, then move the selector switch to lower range positions until the strongest indication is obtained on the meter display without over-ranging it.

If the meter indication is "backwards," (left motion on analog needle, or negative reading on a digital display), then reverse the test probe connections and try again.

When the ammeter indicates a normal reading (not "backwards"), electrons are entering the black test lead and exiting the red. This is how you determine direction of current using a meter.

How to Identify Leaks of Current in Boats and Cars

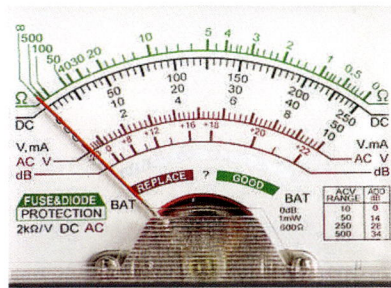

Reading Amps

Chapter 7: DC Panel

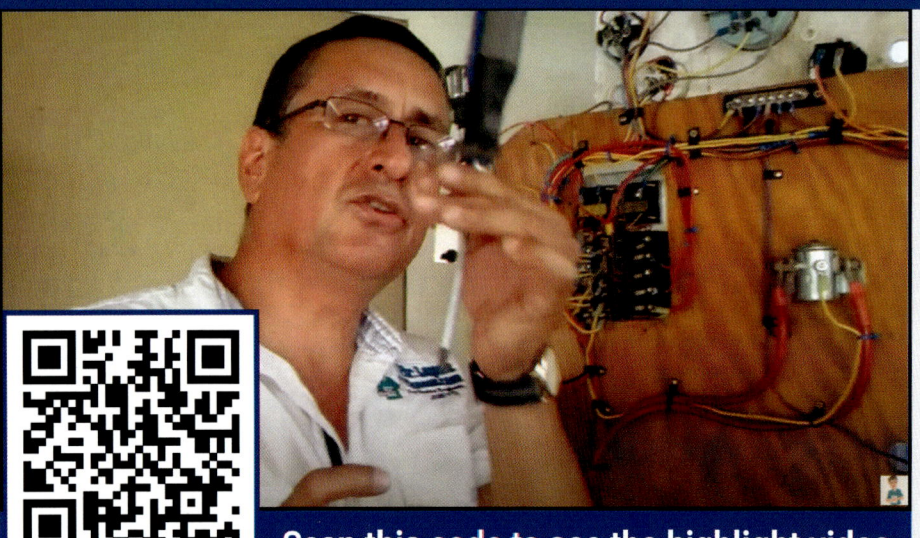

Video Episode 2: Wiring a boat battery bank & DC Panel

In this video you will learn the procedure step by step to do the DC wiring in a pleasure yacht. Starting on the battery banks configuration, then the installation of battery switch selectors and A.C.R and finally the selection and configuration of the main DC Panel

Scan this code to see the highlight video

Follow me

The Voltmeter

A voltmeter is an instrument used for measuring electrical potential difference between two points in an electric circuit.

Analog Voltmeter

Analog voltmeters move a pointer across a scale in proportion to the voltage of the circuit.
When turning the voltmeter on for the first time, note pointer movement. If pointer does not leave the pointer stop pin, the gauge is probably connected backwards. If so, reverse "I" and "G" connections.

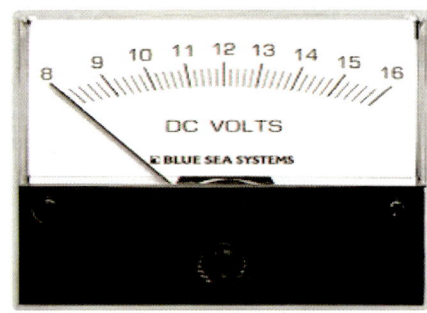 

Chapter 7: DC Panel

Digital Voltmeter

Digital voltmeters give a numerical display of voltage by use of an analog to digital converter.
A voltmeter will not indicate battery condition.

Voltmeter accuracy is affected by many factors, including temperature and supply voltage variations.
To ensure that a digital voltmeter's reading is within the manufacturer's specified tolerances, they should be periodically calibrated against a voltage standard such as the Weston Cell.

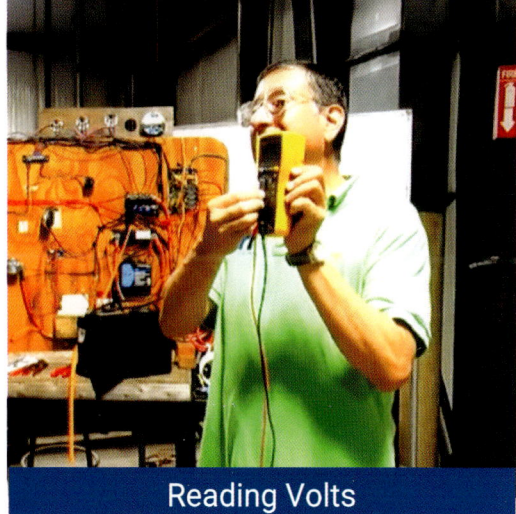

Reading Volts

Marine DC Panel - Master Class

CHAPTER 8

Marine Dashboard

TOPICS

8.1 The ignition Switch	101
8.2 The Fuel Gauge	102
8.3 The Coolant Temp Gauge	103
8.4 Types of Temperature Sensors	104
8.5 Troubleshooting Gauges	105
8.6 Analog and Digital Tachometers	106
8.7 Wiring the Dashboard in a Boat	107

Video Episode 1: Marine Dashboard Configuration

Scan this code to see the highlight video

In this video you will learn the procedure to install and diagnosis the gauges on a boat dash board panel . the video explain how the gauge - the sender and the sensors are integrated in the instruments panel board. The last part of the video is a troubleshoot procedure on the panel.

Follow me

Chapter 8: Marine Dashboard

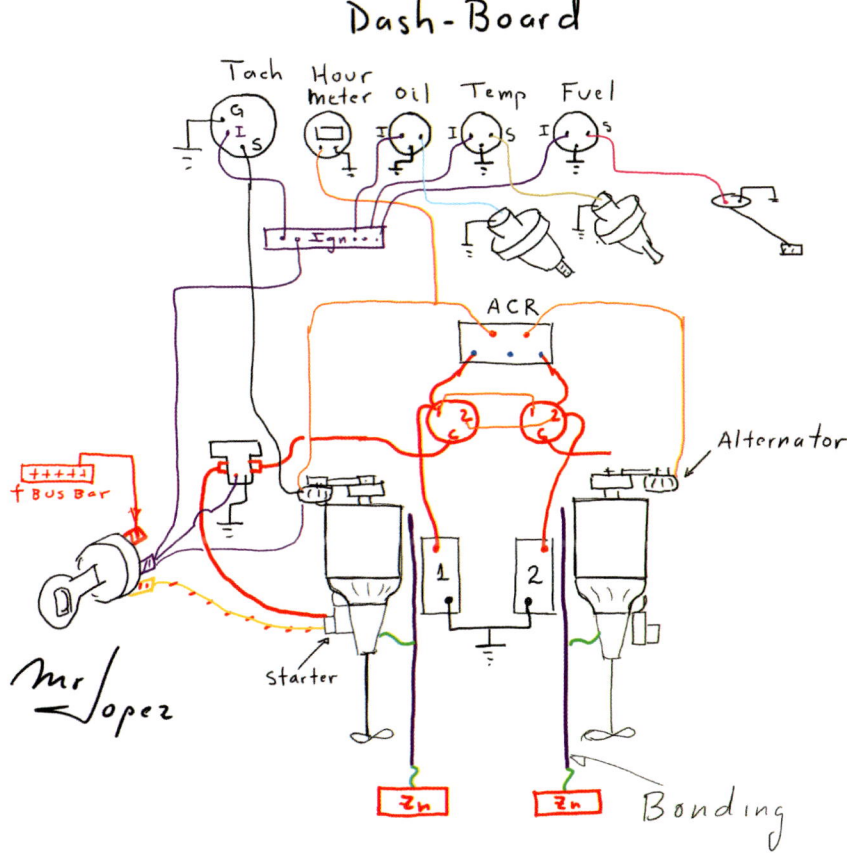

The Ignition Switch

Usually is located on the console. In its motion has three positions, **battery (B)**, **ignition (I)** and **starter (S)**.

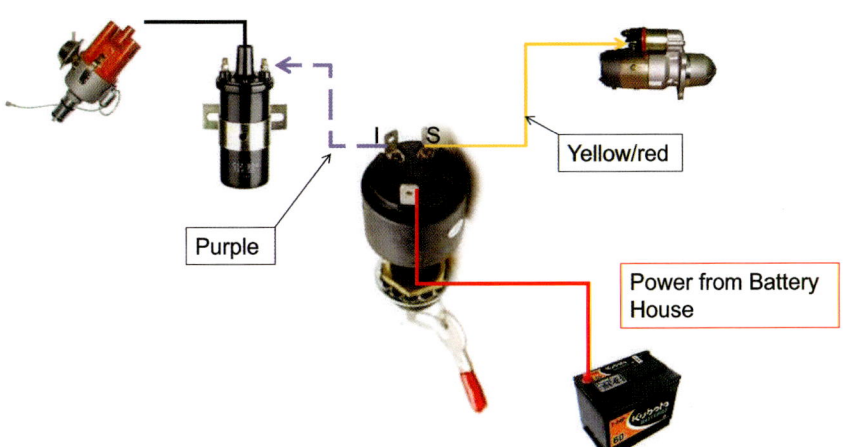

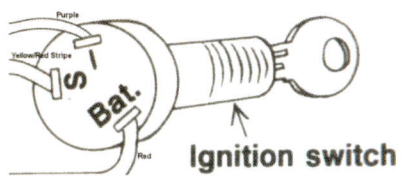

The **battery (B)** terminal receive the power from the positive bus bar. From the **ignition (I)** terminal a signal is send to the start relay and From the **starter (S)** terminal a signal is send to the starter solenoid.

The Fuel Gauge

The sensing unit usually uses a float connected to a Potentiometer. As the tank empties, the float drops and slides a moving contact along the resistor, increasing its resistance.

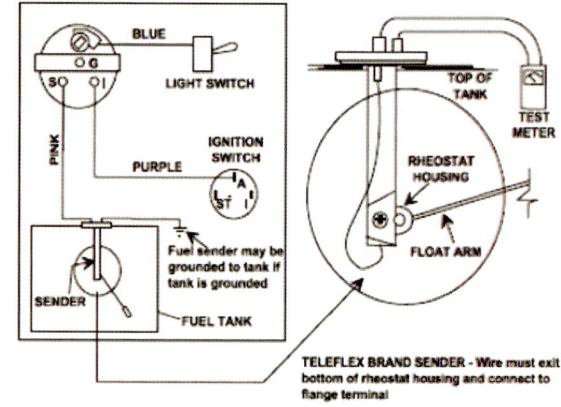

Testing Fuel Gauge

Connect "hot" wire to the "I" terminal and ground wire to "G" terminal. Remove sender (usually pink) wire from back of gauge. Gauge should read below "EMPTY". Next, add a short wire from the gauge's "S" (sender) terminal to ground. Gauge should read above "FULL." If the pointer sweeps back and forth, gauge is OK.

Testing the Sensor

The Float sensor can be tested by checking its resistance with a volt/ohm test meter. Move float arm by hand. Approximate values: Empty = 240 ohms, 1/2 = 103 ohms, Full = 33 ohms.

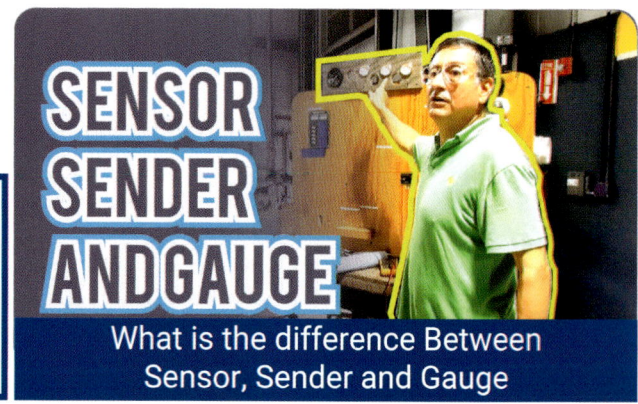

What is the difference Between Sensor, Sender and Gauge

Some Recommendations

Gauge will not operate accurately from more than one sensor at a time. Some installations use a switch to connect one gauge to various tanks, one at a time.
Sender will not operate in water tanks. Rheostat will become electrically "open".
If sender is "open" (infinite resistance) gauge will read below empty. If sender is shorted. (0 resistance) gauge will read above "FULL".

The Coolant Temperature Gauge

Temperature systems used on boats and other vehicles give a general indication of the temperature in which the sender is operating.

±15°F from actual temperature.

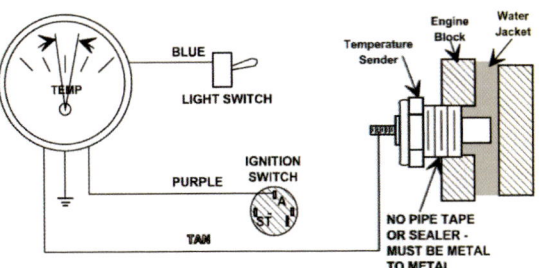

The Temperature Sensor

The sensor (Temp sensor) is also susceptible to "self heating" when electrical current passes through the sender. The self heating causes the sensor to become warmer than the actual temperature of the fluid (the gauge is compensated for this effect at 14 volts)

Testing the Sender

The resistance of the sender can be measured to determine the sender's correct operation. Remove wire to gauge. Connect an ohmmeter to terminal of sensor and to engine block. Approximate values are: 75°F (room temperature) = 600 to 800 ohms; 212°F = 55 Ohm.

Testing the Sensor

If sensor is shorted (0 ohms) gauge will read above 240°F.

If sensor has infinite resistance (Open) gauge will read below 120°F.

If the gauge reads lower than expected, was sealer used on the sensor threads?

Types of Temperature Sensors

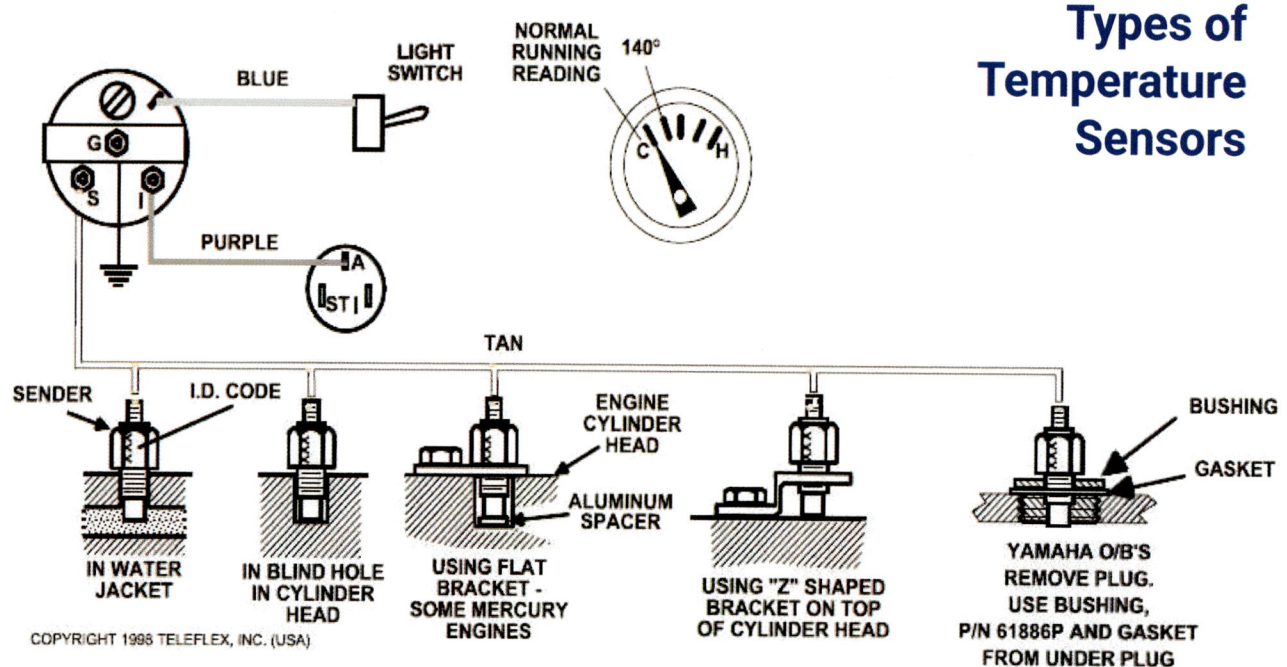

The Oil Pressure Gauge

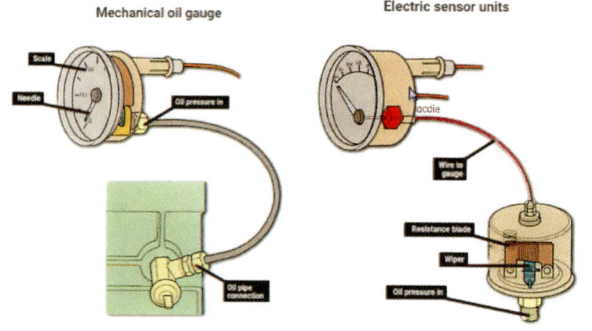

An oil pressure gauge works to ensure that the correct pressure is maintained.
There are two types of oil pressure gauges: the mechanical and electrical kind.
The electrical consists of a gauge itself, an electrical sending unit and a circuit. Inside a mechanical gauge is a spring. The position of this spring is altered by the pressure of the oil indicating rising or falling pressures.

Whereas the pressure readings in an electrical gauge are altered when electrical signals are sent to the gauge passing through a coil altering the needle arrangements indicating the pressure.

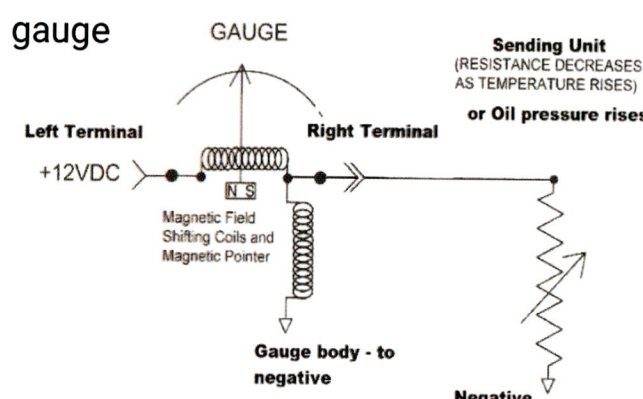

Oil Pressure Gauge Wiring

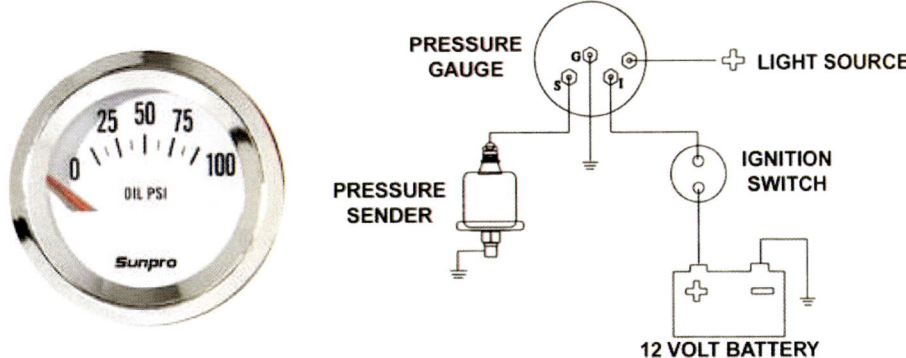

Mechanical Oil Pressure gauge

This Pressure Gauge is bolted directly to the engine block by means of a flexible high pressure pipe.
His only power connection is for the lighting bulb.

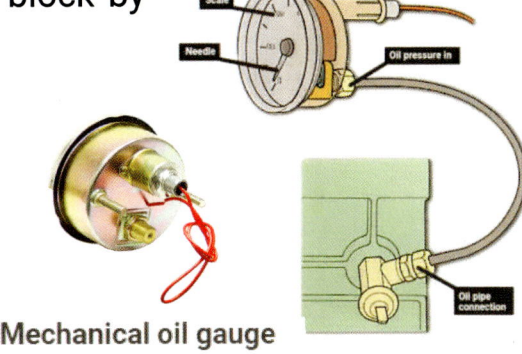

Mechanical oil gauge

Troubleshooting Gauges

Remove the gauge's sender wire. Turn on the power. The pointer of whatever gauge you are checking should be at the position shown in the upper portion of the diagram (Next Slide)
Next, take a short wire and connect to sender terminal and ground terminal (shorting sender terminal to ground). Gauge pointer should be at the position shown in the lower portion of the diagram.

Chapter 8: Marine Dashboard

Analog and Digital Tachometers

Analog Tachometers for Diesel Engines

Analog Tachometers for Diesel Engines Part II

Analog Tachometers Installation Process

Chapter 8: Marine Dashboard

Wiring the Dashboard in a Boat

Wiring a Boat Dash Board

Wiring a Boat Dash Board Part II

Chapter 8: Marine Dashboard

Engine Harness Wiring Diesel Engine

Scan this code to learn how to build the harness

You will find more Posters at: www.mrlopezclasses.com - www.mtttedu.org

CHAPTER 9

AC Current

TOPICS

9.0 AC Current Production	110
9.1 The Phase Angle	112
9.2 The Sine Wave	113
9.3 The Frequency	114
9.4 Frequency Vs RPM	115
9.5 American Power vs European Power	117
9.6 The Amplitude	117
9.7 Wavelength	117

Video Episode 1: Sources of AC current

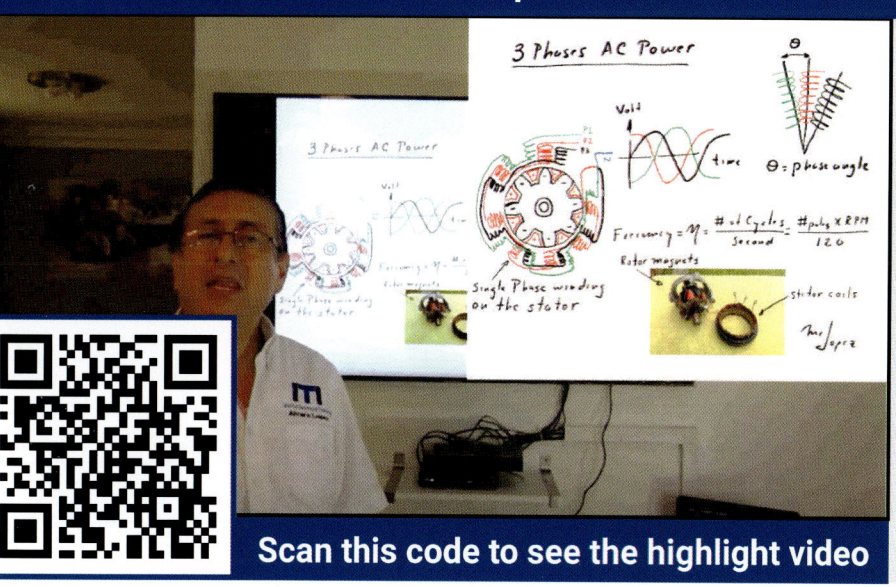

Scan this code to see the highlight video

In this channel we will learn about the sources of AC current in a typical pleasure boat or mega-yacht. We are going to study the wavelength of the AC current, the frequency 50 Hz and 60 Hz . the amplitude and the production of single phase, double phase and three phases .

Follow me

AC Current

If a machine is constructed to rotate a magnetic field around a set of stationary wire coils with the turning of a shaft, AC voltage will be produced across the wire coils as that shaft is rotated.

Generators, Hydroelectric power plants and Nuclear power plants, are sources of AC power.

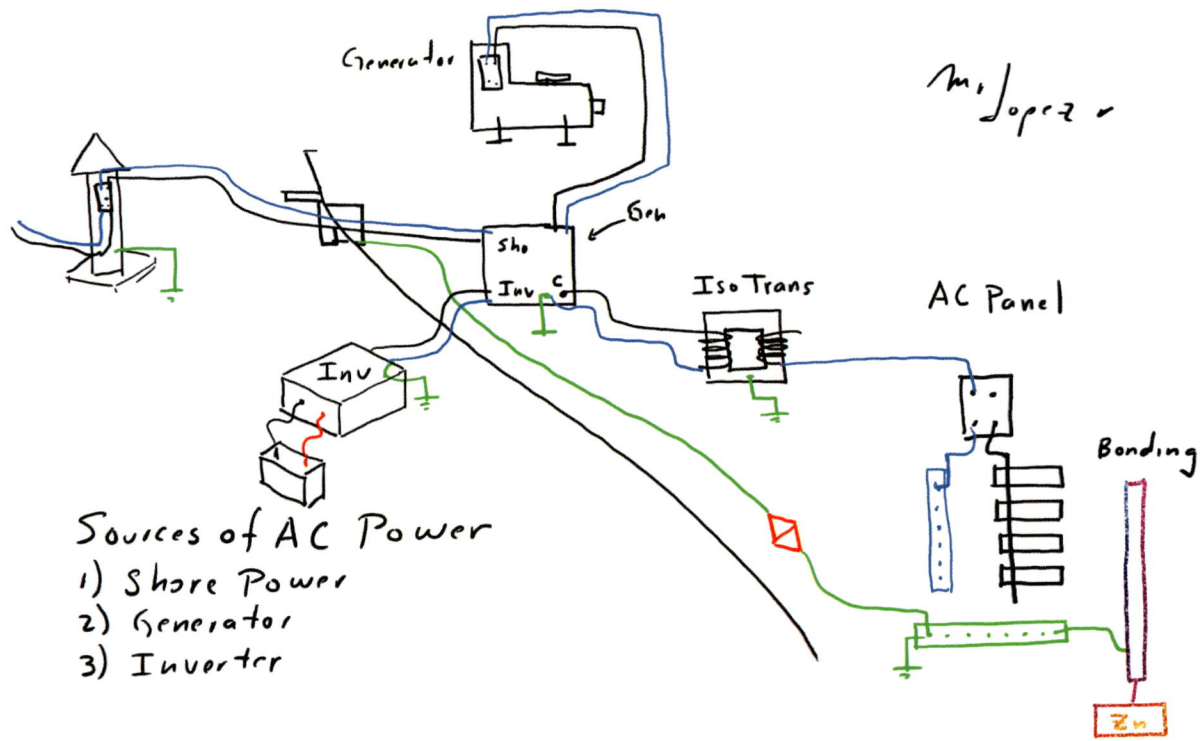

Sources of AC Power
1) Shore Power
2) Generator
3) Inverter

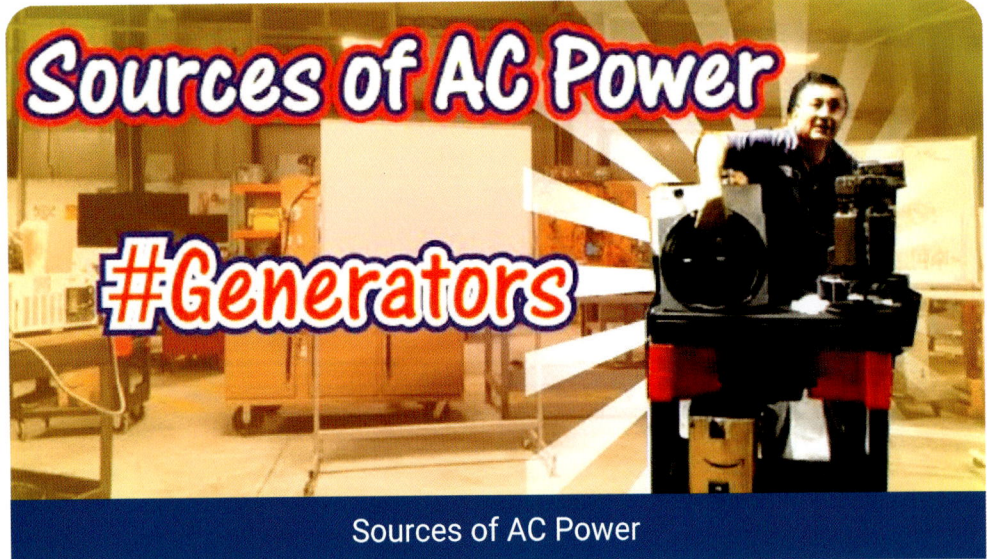

Sources of AC Power

AC Wavelength

The **Amplitude** is a Function of: The thickness of the wire coil and the numbers of coil turns.

The **Frequency** is directly proportional to the engine RPM more RPM then more Frequency.

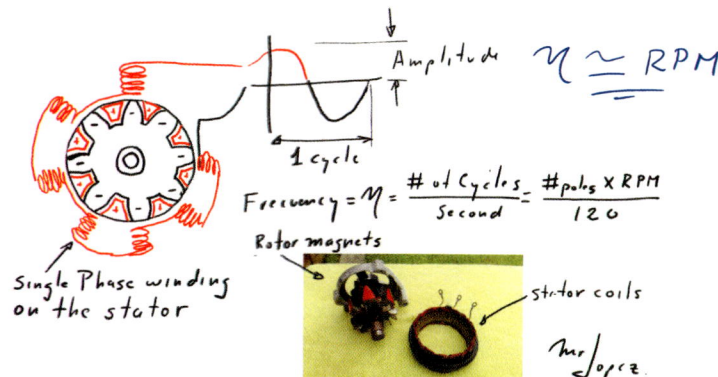

AC Current

The constant motion of the armature creates the electrical current, and its changing position relative to the magnetic field causes the "alternation".

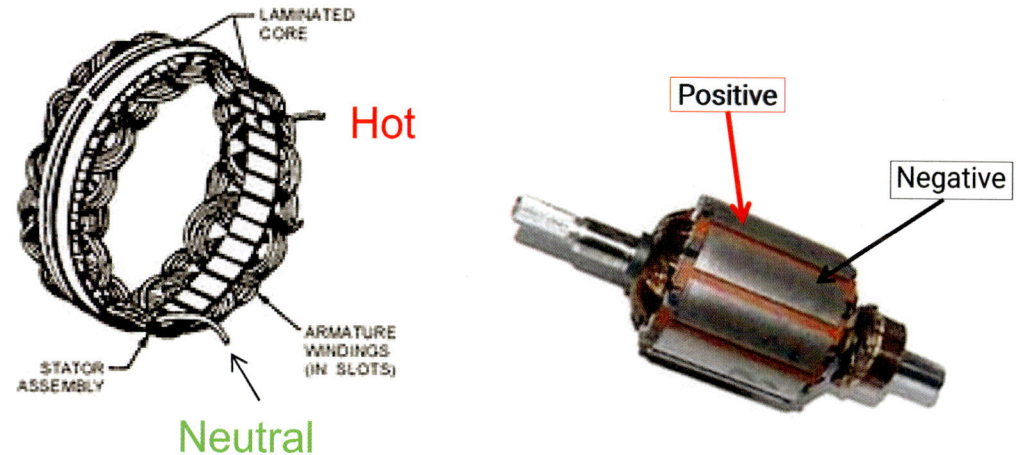

Alternating Current (AC) is a form of electrical current where the polarity and direction reverses in regular cycles.

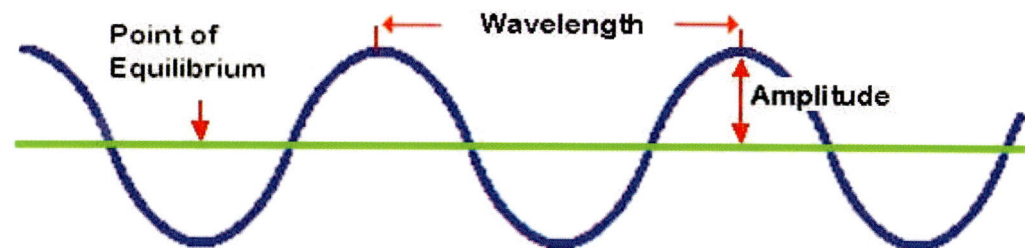

AC Current

The basic AC generator uses a coiled conductive wire called an armature, which is rotated in the field created between **two magnets.**

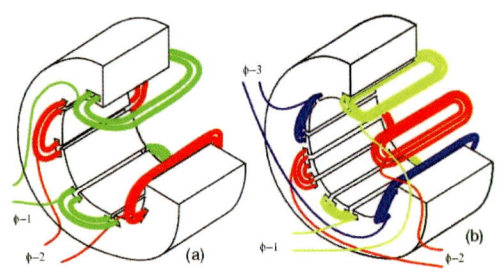

The Phase Angle

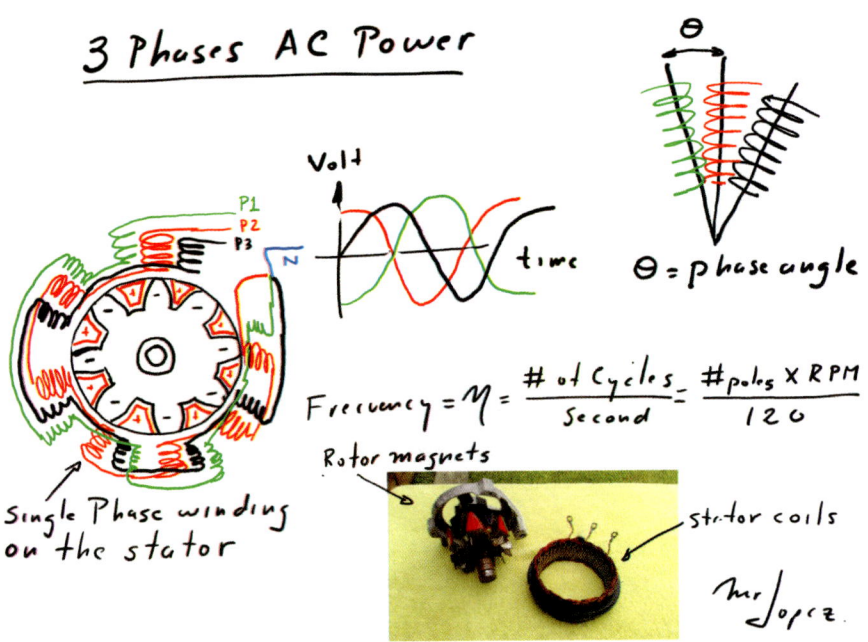

Depending of the amount of windings at the stator, One, Two or Three phases can be generated.

During the winding procedure, each phase is separated some degrees with respect to the next one. This separation is called 'The Phase Angle".

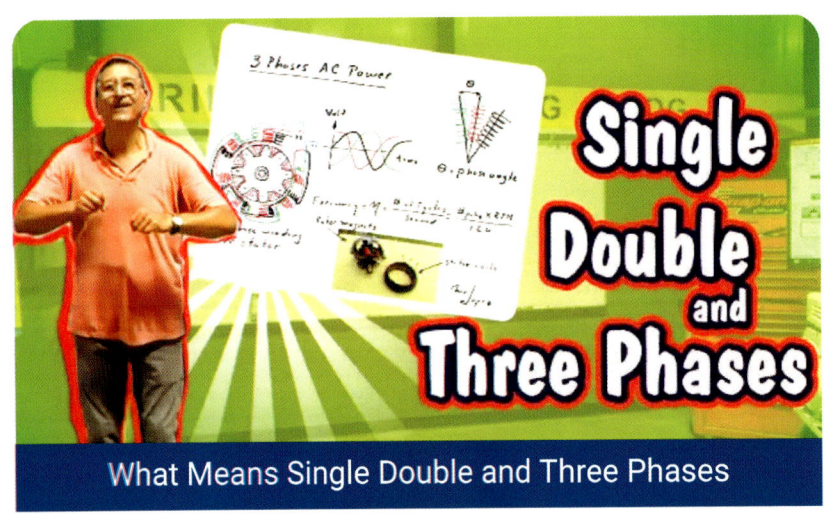

What Means Single Double and Three Phases

The Sine-wave

When an alternator produces AC voltage, the voltage switches polarity over time, but does so in a very particular manner.

When graphed over time, the "wave" traced by this voltage of alternating polarity from an alternator takes on a distinct shape, known as a sinewave.

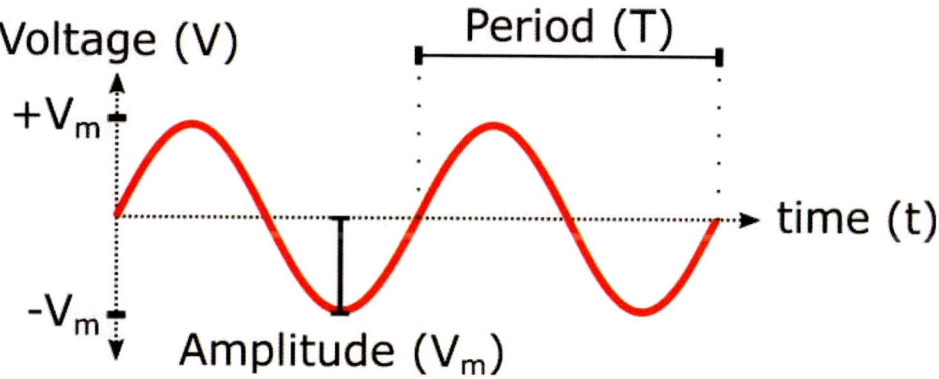

Wave Shape

In certain applications, different waveforms are used, such as triangular or square waves.

Audio and radio signals carried on electrical wires are also examples of alternating current. In these applications, an important goal is often the recovery of information encoded (or modulated) onto the AC signal.

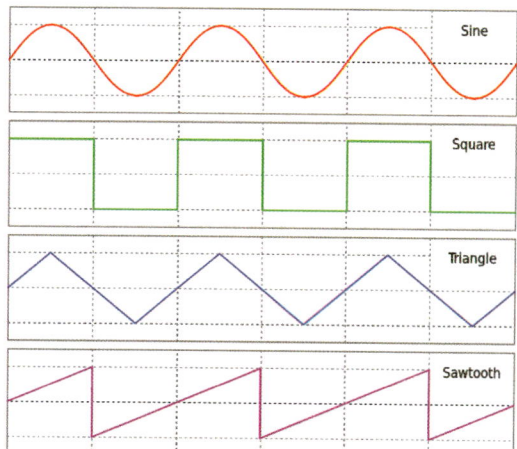

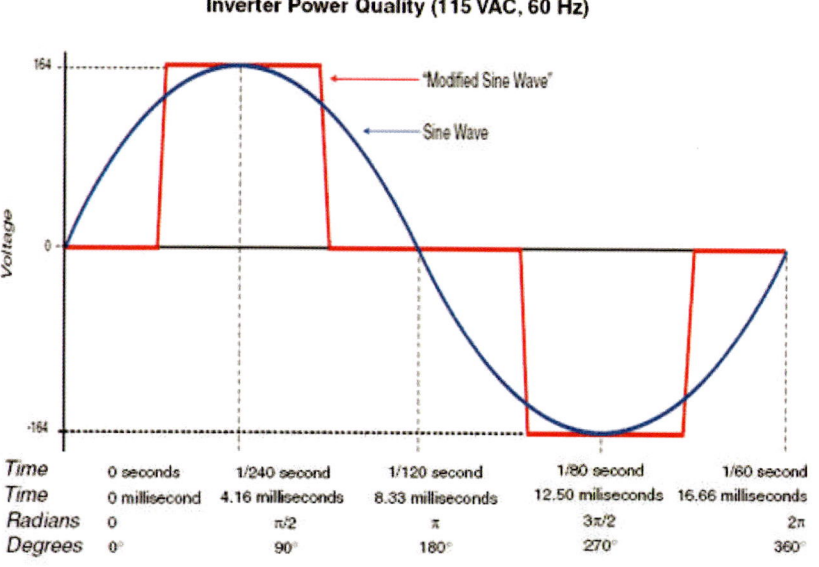

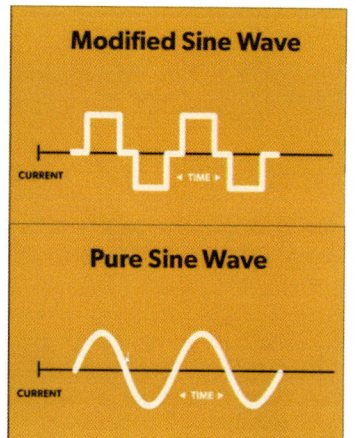

The Frequency

Imagine two adjacent portions of the rotor (**Pos**. & **Neg**) passing in front of the coil (Pole) during the rotation of the rotor.

When the magnets (**+**) and (**-**) are passing in front of the Pole 1, a complete cycle is produced.

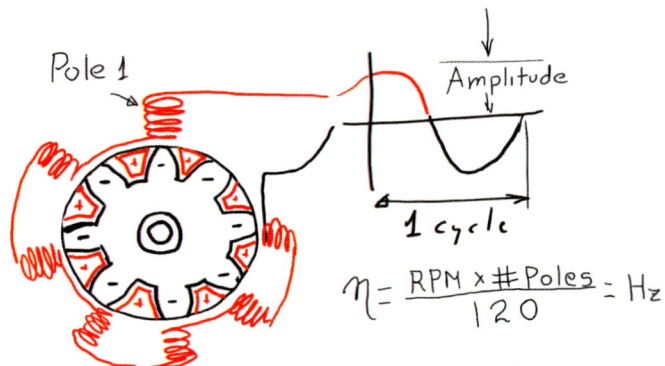

Frequency Calculator

The frequency of the generated voltage is dependent on the number of field poles and the speed at which the generator is operated.

$$\eta = f = \frac{NP}{120} = Hz$$

- f = Frequency (Hz) ; N = rpm; P = # of Poles
120 = Conversion from min to sec and from poles to pole pairs.

$$\frac{60 \text{ seconds}}{1 \text{ minute}} \times \frac{2 \text{ poles}}{\text{pole pair}}$$

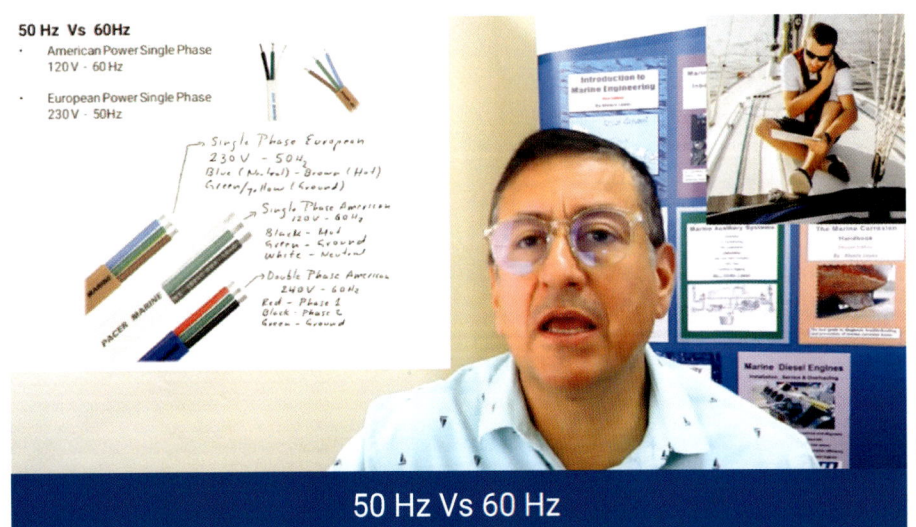

50 Hz Vs 60 Hz

Chapter 9: AC Current

Frequency Vs Engine RPM

The frequency of the generated voltage is dependent on the number of field poles and the speed at which the generator is operated. Frequency, measured in Hertz (Hz), is the number of complete cycles per second in alternating current direction.

$$\eta = \text{Frequency (Hz)} = \frac{\text{RPM} \times \text{Poles}}{120}$$

The frequency is directly proportional to the number of revolutions of the motor. More engine RPM, more frequency

As we will study later, never try to increase or decrease the output voltage changing the engine RPM because the frequency is altered and consequently the efficiency decreases.

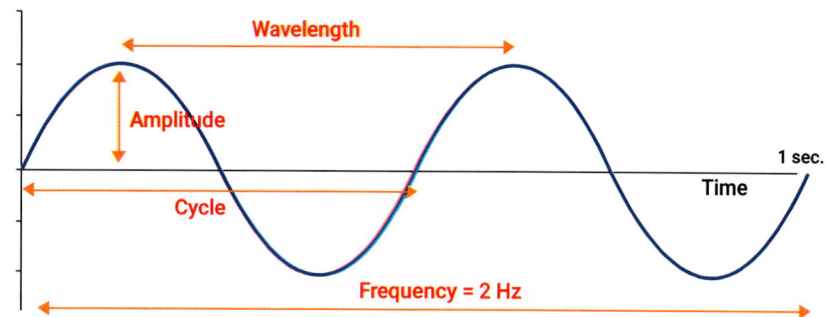

The Frequency

The frequency of the electrical system varies by country; most electric power is generated at either 50 or 60 Hz. Some countries have a mixture of 50 Hz and 60 Hz supplies.
Frequency is always measured and expressed in hertz.

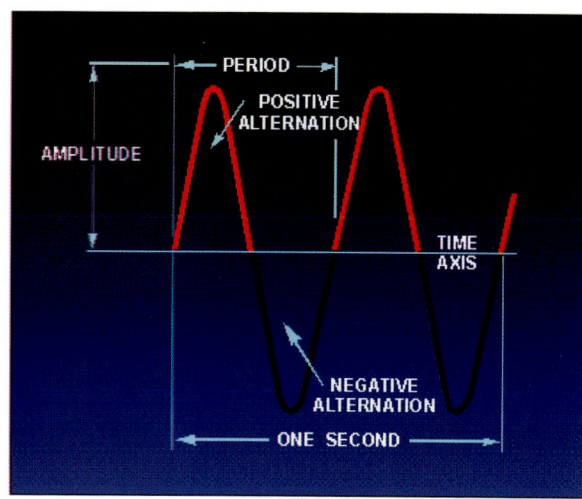

Low Frequency

A low frequency eases the design of low-speed electric motors, particularly for hoisting, crushing and rolling applications, and commutator-type traction motors for applications such as railways, but also causes a noticeable flicker in incandescent lighting and an objectionable flicker in fluorescent lamps.

High Frequency

Off-shore, military, textile industry, marine, computer mainframe, aircraft, and spacecraft applications sometimes use 400 Hz, for benefits of reduced weight of apparatus or higher motor speeds.

How to Read frequency

The Period

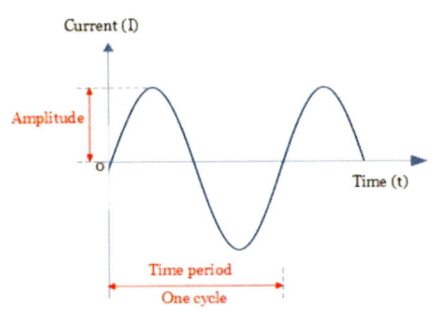

The time required to complete one cycle of a waveform is called the PERIOD of the wave.
In the picture, the period is one-half second.

Time / Frequency

The relationship between **time (t)** and **frequency (f)** is indicated by the formulas.

$$t = \frac{1}{f} \quad \text{and} \quad f = \frac{1}{t}$$

where t = period in seconds an
f = frequency in hertz

American Power 60 Hz vs European Power 50 Hz

The process to Calibrate the Frequency

The Amplitude

The distance from zero to the maximum value of each alternation is called the AMPLITUDE. The amplitude of the positive alternation and the amplitude of the negative alternation are the same.

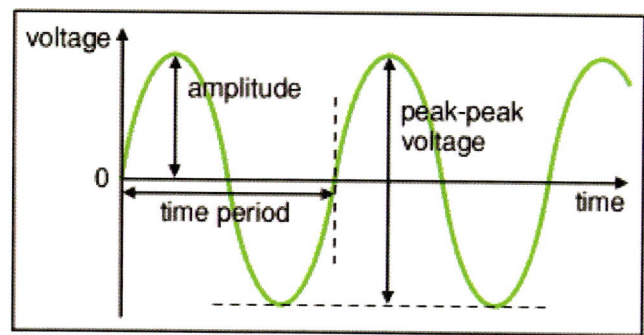

Wavelength

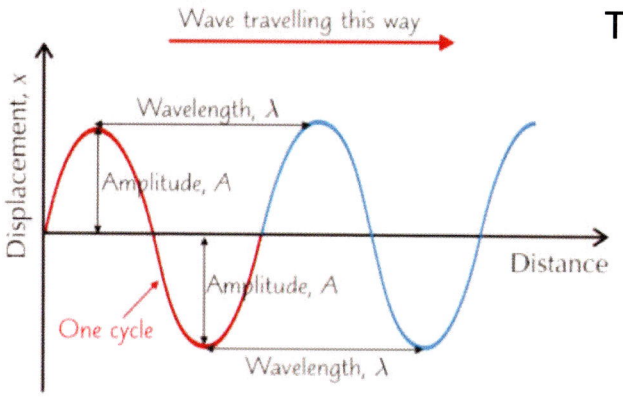

The time it takes for a sine wave to complete one cycle is defined as the period of the waveform. The distance traveled by the sine wave during this period is referred to as WAVELENGTH. Wavelength, indicated by the symbol & lambda.

CHAPTER 10

Bonding & Lightning

TOPICS

10.0 Bonding & Grounding	119
10.1 Cathodic Protection	121
10.2 Common Bonding Conductor	122
10.3 AC Ground Faults	123
10.4 Galvanic Isolator	124
10.5 Isolator Transformer	127
10.6 Lightning Protection	128
10.7 Galvanic Corrosion	132
10.8 Electrolytic Corrosion	133

Video Episode 1: Bonding System

Scan this code to see the highlight video

In this video I explain the Bonding System and how the bonding connection in a boat is critical in order to keep the boat away from corrosion.
Also the video explain how the reverse polarity condition is critical in terms of corrosion.

Follow me

Bonding and Grounding

"Bonding" refers to a common connection joining metal components to:

- Provide a low impedance ground fault path to trip a circuit protection device.
- Prevent dangerous voltages from appearing on metal objects.
- Provide a path for galvanic and DC stray currents.

"Grounding" is defined as a common connection to Earth for the purpose of:

- Lightning discharge
- System voltage stabilization
- Reducing static and RF interference

Bonding or Grounding ?

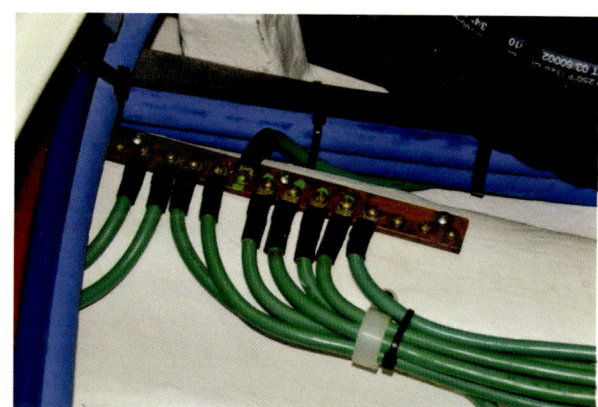

Keep in mind that these terms are often used interchangeably in the current lexicon
We will be focusing primarily on "Bonding" in its classic sense but may also refer to some aspects of this as "Grounding" (e.g. the green wire is called the ground wire, but it's actually a bonding wire by definition in most cases).

Bonding, Grounding - Current Carrying Conductor

DC (Negative) Bus Bar Vs AC (Ground) Bus Bar

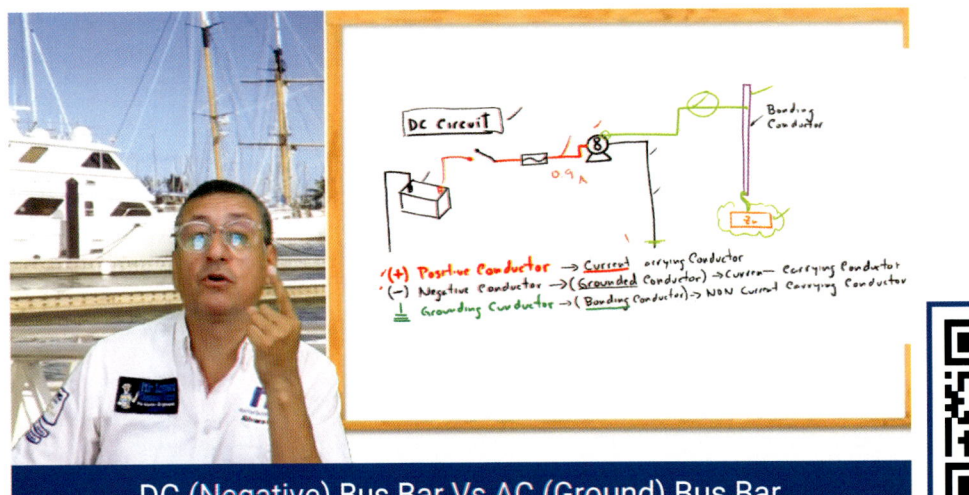
DC (Negative) Bus Bar Vs AC (Ground) Bus Bar

What is Bonding?

Is the electrical connection of the exposed, metallic noncurrent carrying parts to the ground side of the direct-current system.

BONDING Conductors are normally noncurrent-carrying conductors used to connect the noncurrent-carrying metal parts of a boat and the noncurrent-carrying parts of direct-current devices on the boat to the boat's bonding systems.

Common Bonding Conductor is an electrical conductor, usually running fore-and –aft along the boat's center line, to which all equipment bonding conductors are connected.

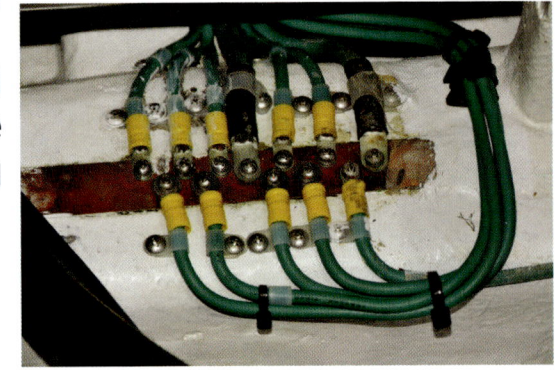

GROUND is a surface or mass at the potential of the earth's surface, established by a conducting connections (intentional or accidental) with the earth.
In practical terms, ground is the potential or voltage , of the water in which the boat is immersed.
Note: The voltage in the water may vary slightly due to stray currents.
The purpose of the bonding system is to force this voltage to be as uniform as possible through the use of low resistance conductors and connections.

Bonding and Cathodic Protection

Cathodic protection is the practice of connecting all hardware in contact with the water to an external sacrificial anode(s) also called Zinc's.
This is usually done by creating a common buss where we then connect all the underwater metals from inside the boat.

Things usually connected to this buss would include things like through hull fittings, engines, generators, rudder shaft's, and anything else that is in contact with the outside water.

Extra zinc's can also be fitted onto outdrives and outboard engines along with rudders, skegs and keels.
Bonding is the practice of connecting all major metal components on a boat that may be required to safely conduct electricity to the boats designated ground.

That buss is then connected to an external anode via one of the bolts attaching the anode to the outside of the hull.

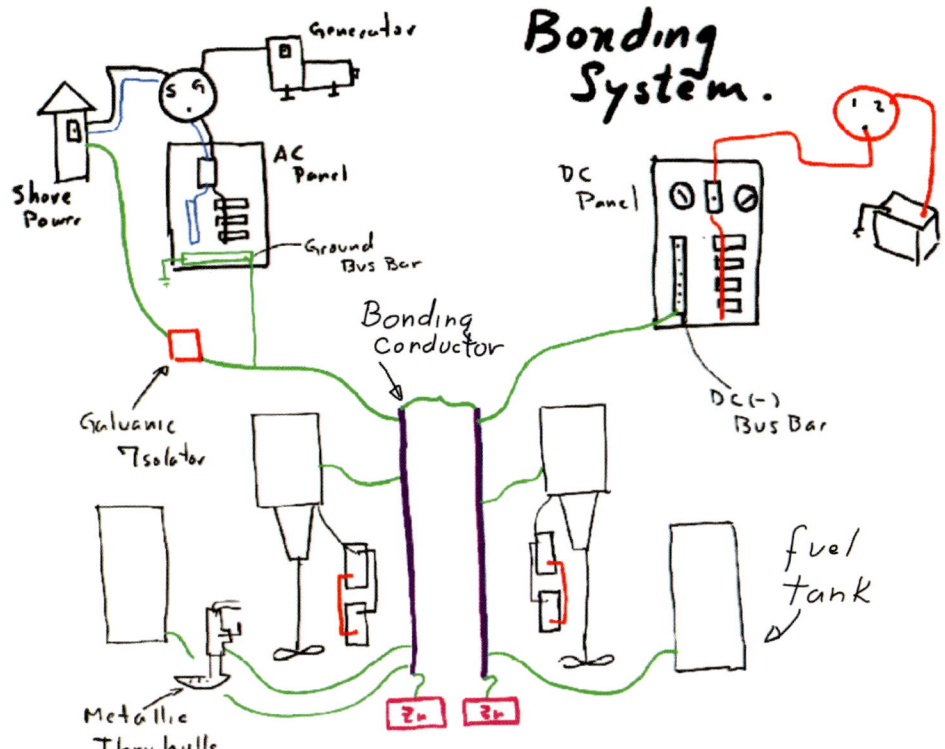

Common-Bonding Conductor

Bonding conductor

The large common-bonding conductor should be bare or green.
If wire, it may consist of bare, stranded, tinned copper or insulated stranded wire of minimum size 8 AWG
If solid, it may be uninsulated copper, bronze strip, or copper pipe at least .030 in. thick and 0.5 in. wide.
The copper, bronze strip or pipe may be drilled and tapped, provided it is thick enough to provide at least three full threads for terminal screws.

Equipment to be connected to the boat's bonding system includes.
- Engines and Transmissions.
- Propellers and Shafts.
- Electronics Cabinets
- Metal Cabinets and control boxes.
- Metal fuel and water tanks, and electrical fuel pumps and valves.
- Metal battery boxes
- Metal conduit or armoring
- Trim Tabs flaps

As defined by ABYC Standard E-9.
DC Grounding Conductors are normally noncurrent-carrying conductors. They are used to connect metallic noncurrent-carrying parts of DC devices to the engine negative terminal or its bus.
DC Grounded conductors are current-carrying conductors connected to the side (usually negative) of the source.
Ground is established by a conducting connection with the earth, including any conductive part of the wetted surface of a hull.

AC Ground Fault Effects

If a ground fault causes a metal or metal-cased component to rise to dangerous potentials (again, from loss or poor green ground wire), and a person touches this object along with another object at a lower potential, the person bridges the gap and receives a shock from the difference in the 2 potentials.

AC Ground Fault Effects

1 ma	Tingle
15 ma	Paralysis/Drowning
60 ma	Heart Failure
300 ma	Combustion
10 ma	= .01 Amps

NFPA 70 (National Electric Code) considers that potentials as low as 30vac are considered dangerous. This situation can produce lethal currents in the body.

Broken Ground

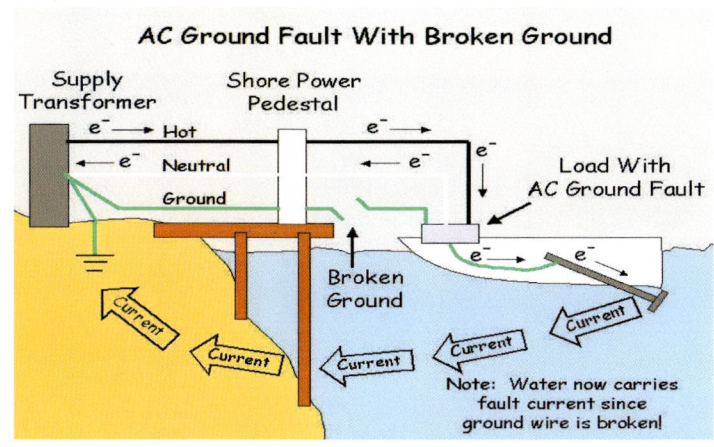

If the ground line is broken, the static electricity stored in the case of equipment's and appliances try to return to the source of the power through the shaft and propeller, increasing the electrolysis on the exposed metals.

Testing Shore Power & Boat Grounds

When two or more boats are connected to shore power, one side of the necessary circuit required to form a galvanic couple is provided by the AC green grounding wire, which is also connected to the boat ground system, engine, and underwater hardware.

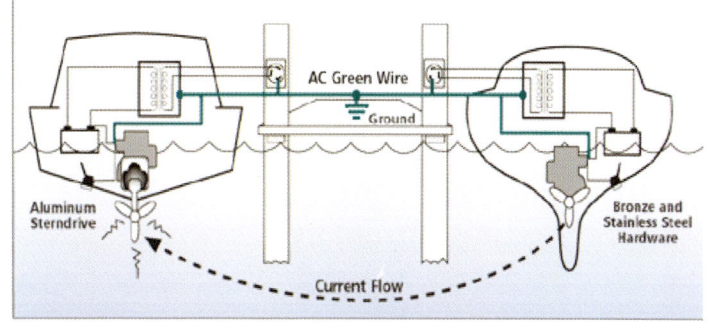

Chapter 10: Bonding & Lightning

Galvanic Isolator

The Galvanic Isolator blocks electrolysis currents from flowing in the ground conductor of your shore power hookup. It provides approximately 1.2 volts of isolation to isolate electrolytic voltages from the dock but yet pass safety currents to ground in the event of a short circuit, or power leakage on your boat.

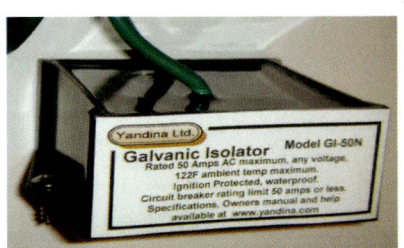
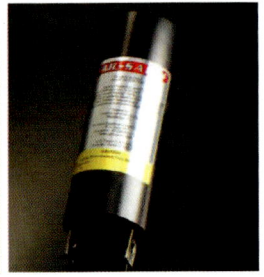

Video Episode 2: How to check and install a galvanic Isolator

In this video you will learn the procedure to install and check a galvanic isolator and how the reverse polarity condition affects your boat. This is an important reference material for boat surveyors, boat inspectors and marina managers.

Follow me

Scan this code to see the highlight video

What is Galvanic Isolator ?

A galvanic isolator is a device used to block low voltage DC currents coming on board your boat on the shore power ground wire. These currents could cause corrosion to your underwater metals; through hulls, propeller, shaft etc.

Boats in a marina plugged into shore power all act as a giant battery. They are all connected together by the green shore power ground wire, which is (or should be) connected to their DC grounds, engine block, and bonded underwater metals. If the boats are in salt water, then that forms an electrolyte and the dissimilar metals connected together act as a battery, causing corrosion.

Galvanic Isolator Location

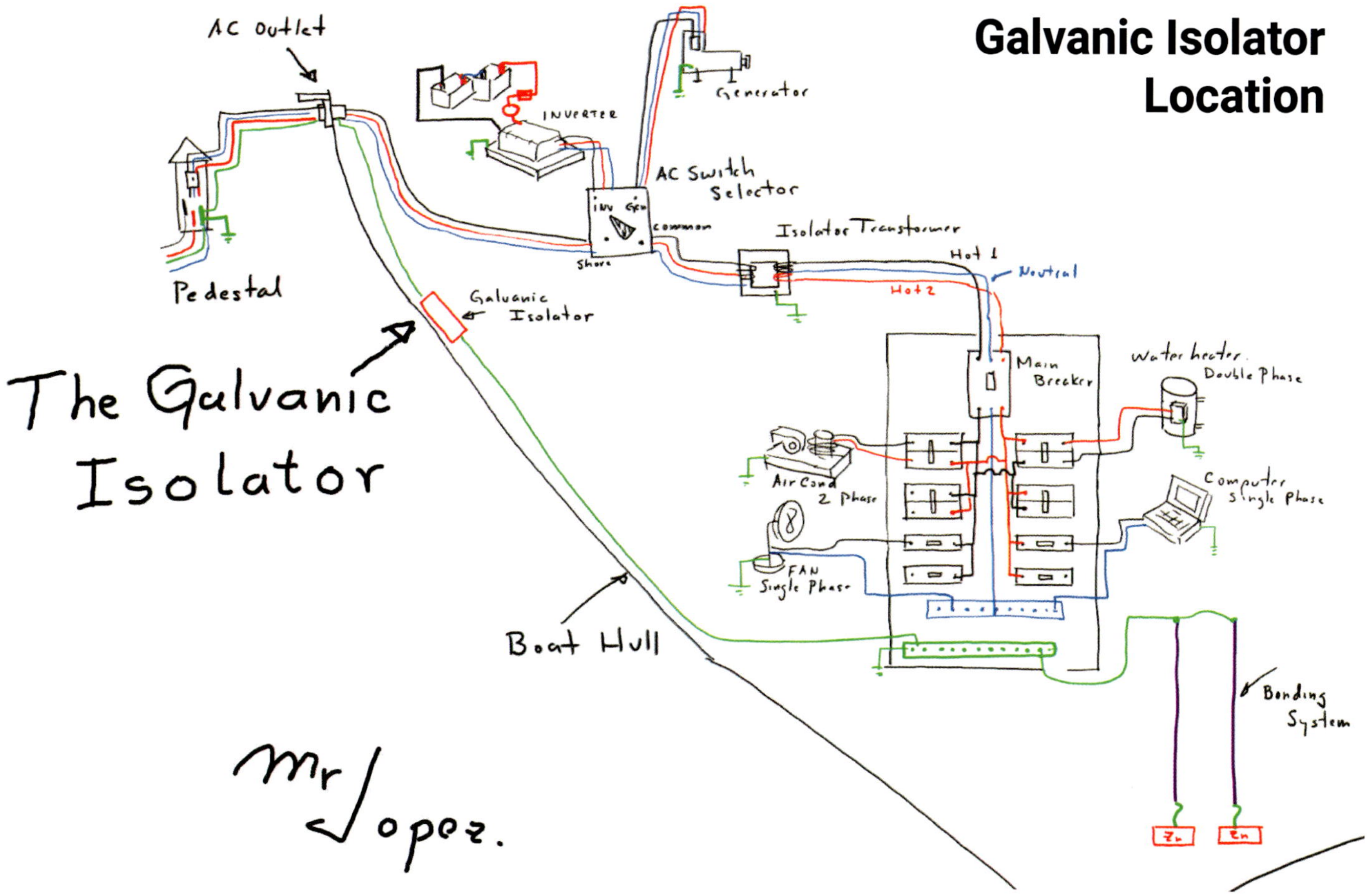

The Galvanic Isolator

Galvanic Isolator Functions

Normally no AC current is carried on the shore power ground wire, but it has to be able to carry the full load of the circuit in the event of a fault.
Therefore it is important to have a good quality unit that will not overheat when required to carry the rated load. Some heat will be generated by the voltage drop and the unit must be able to withstand this.

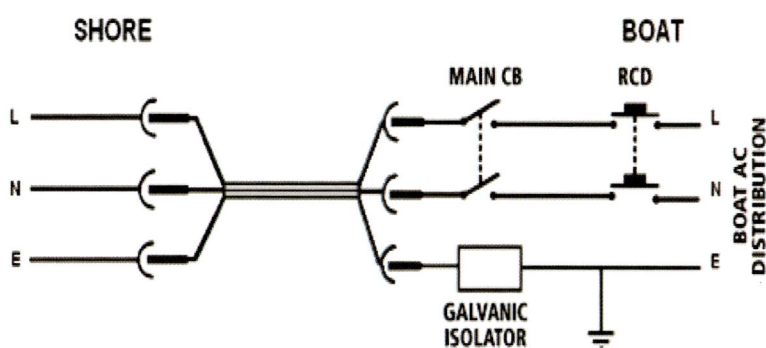

Galvanic Isolators

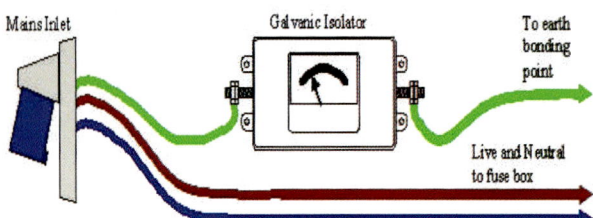

If you connect to shore power in a marina you should invest in a galvanic isolator. As soon as you plug in you connect yourself to your neighboring boats through the ground wire in your shore power cord.

This creates a galvanic circuit between you and the other boats(not good if your neighbors' have bad stray electrical currents on their boats, it will eat up your anodes fast).

The galvanic isolator interrupts low galvanic voltage/current but will connect larger amounts in the case or an appliance short that needs to be grounded.

But before you storm down to your marina demanding retribution, you should know that the marina wiring plays only an indirect role in a problem born of correct wiring practices. The solution is a galvanic isolator.

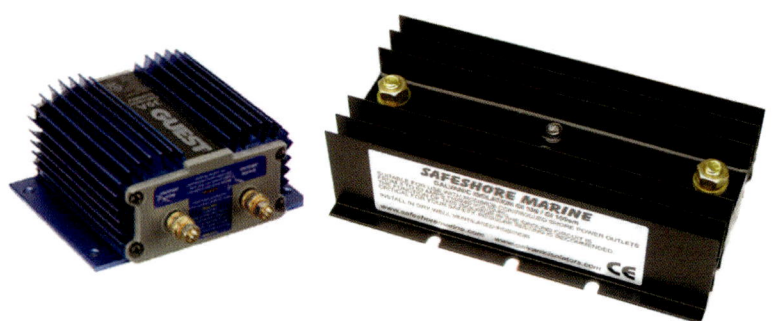

Testing a Galvanic Isolator

Disconnect one lead of the isolator so that you are testing it only.

Get a digital multimeter set to the diode test function. Put one lead on one side of the isolator and the other lead on the other side.

As the capacitor starts to conduct current the reading should rise to approximately 0.9 volts.

Remove the test leads, short the two wires of the isolator together to discharge the capacitor and repeat the test with the test leads reversed. You should get the same answer.

Interpreting the readings

If the reading is instantly 0.9 volts, then the capacitor is defective or there is no capacitor.

If a voltage of 0.45 volts is observed one of the diodes is shorted.

If there is a reading of **0** volts then both diodes could be shorted.

If there is a reading in excess of **0.9** volts then one or both diodes are open (not conducting) in which case you should stop the test before the capacitor reaches **2.0** volts or you will damage it.

Video Episode 3 Isolator Transformer vs Galvanic Isolator

In this video you will learn about the cranking systems for inboard and outboard gasoline and diesel engines. The last part of the video is related with the diagnosis process.

Follow me

Scan this code to see the highlight video

Isolator transformer

Isolation Transformers have primary and secondary windings that are physically separated from each other. Sometimes isolation transformers are referred to as "insulated"

Isolator Transformer

Grounding Path

Contrary to popular belief, electricity takes all paths back to its source, not just the path of least resistance.

This presents a clear hazard to swimmers, and those aboard may also be at risk, depending upon the nature of the fault.

Therefore, it is vitally important that the **connection between the AC safety grounding wire and the neutral wire occur only at the power source !!**

Lightning protection

The combination of vertically moving water droplets and air currents results in the buildup of large quantities of oppositely charged particles within clouds and between clouds and the ground.

The electrical potential differences between charges may be as high as **100'000.000** Volts.

By comparison, the voltage on the power lines running along a street is **12.000 Volts.**

The base of a cloud becomes negatively charged. Since opposite charges attract, the surface of the earth directly beneath the cloud becomes positively charged.

Isolator transformer

In an isolation transformer the output winding will be isolated, or floating from earth ground unless bonded at the time of installation Secondary neutral to ground bonding virtually eliminates common mode noise, providing an isolated neutral-ground reference for sensitive equipment and an inexpensive alternative to the installation of dedicated circuits and site electrical upgrades.

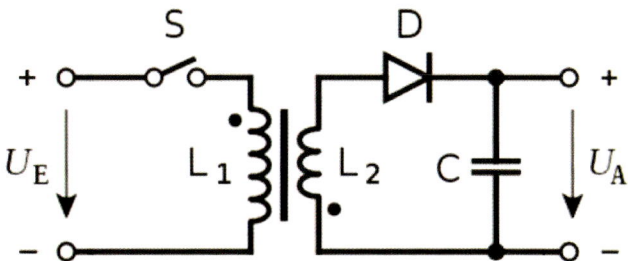

Transformer Isolator Functions

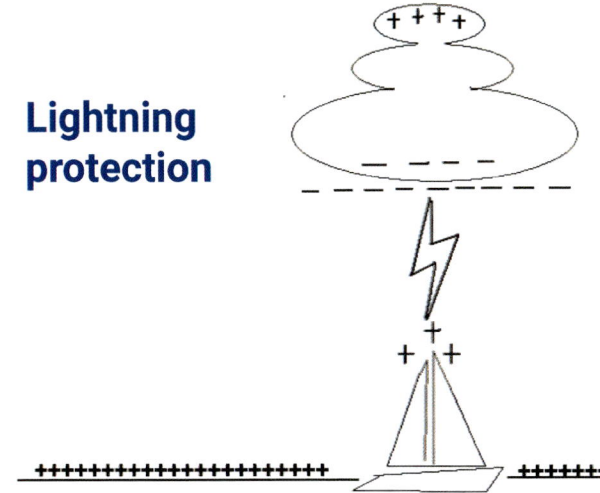

Lightning protection

The primary and secondary windings may be constructed to step-up or step-down the output voltage.

Isolation transformers constructed with Faraday shields, will improve power quality by attenuating higher frequency noise currents.

Isolation transformers offers 100% isolation from the input AC line.

Galvanic Isolator vs Isolator Transformer / Master Class

Lightning protection

The base of a cloud becomes negatively charged. Since opposite charges attract, the surface of the earth directly beneath the cloud becomes positively charged.

It is very difficult to study lightning for many reasons. It occurs very randomly even within ideal conditions and varies greatly in its attributes; a bit akin to snowflakes, no two lightning bolts are alike.

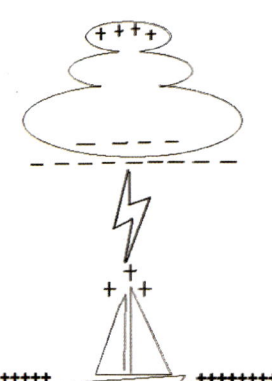

In this video you will learn about the phenomenon of lightning in a boat. How to protect your boat according with the standards recommended by **ABYC** and federal regulations.

Follow me

Lightning Protection for Powerboats

For power boats I recommend having a short mast with a lightning rod attached. If the mast is metal use 4awg wire at the base of the mast to connect through to our grounding source.

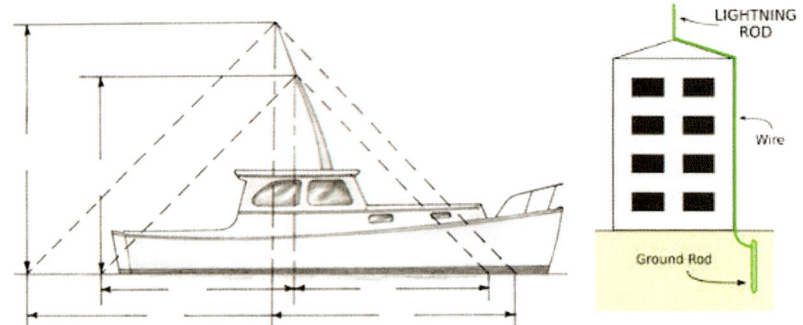

Chapter 10: Bonding & Lightning

Lightning Protection for Powerboats

If you don't have a metal mast you will need to run 4awg wire all the way from the lightning rod to the grounding source. Your engines should also have a 4awg wire going bonding wire going directly to the grounding plate for bonding as well as extra protection against stray lightning current/voltage.

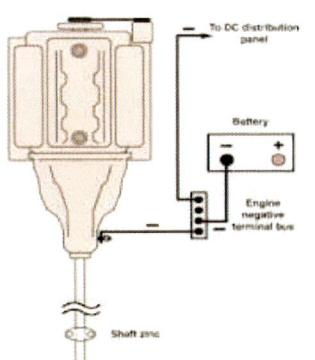

For sailboats with aluminum masts

For sailboats with aluminum masts attaching a 1/4" stainless steel rod shaped to a point with a good connection to the mast will do. The mast will conduct the lightning down.

The top of the lightning rod should be 6" above anything else on the mast and filed to a sharp point. At the base of the mast, we have to connect directly as possible to a good ground ie external keel through a keel bolt or attaching to a metal centerboard.

The top of the lightning rod should be 6" above anything else on the mast and filed to a sharp point. At the base of the mast we have to connect directly as possible to a good ground ie external keel through a keel bolt or attaching to a metal centerboard.
It should be made of corrosion resistant materials such as
 (copper, bronze or monel) and **no less than 1 ft square.**

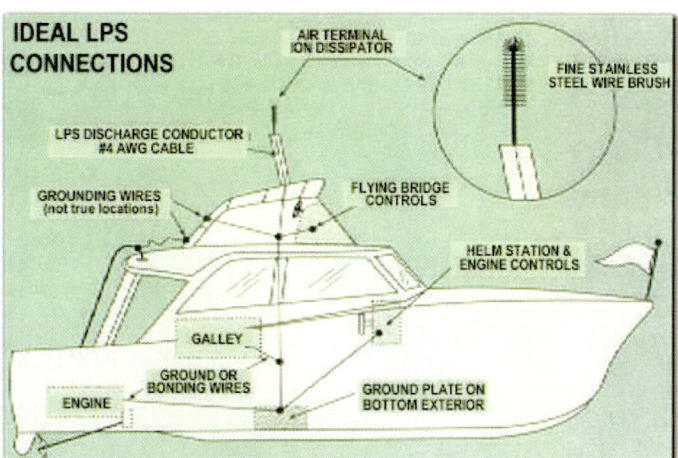

Chapter 10: Bonding & Lightning

For sailboats with aluminum masts

Some of the lightning may try travel down the rigging so we recommend connecting 6 awg cable to the base of each chain plate going directly as possible from there-to-there boats ground. Our bonding circuit is also connected to that ground.

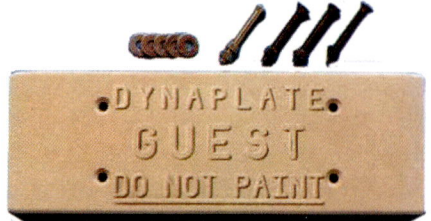

On a sailboat with a wooden mast run 4awg wire from the lightning rod all the way down the mast to the boats main grounding source.

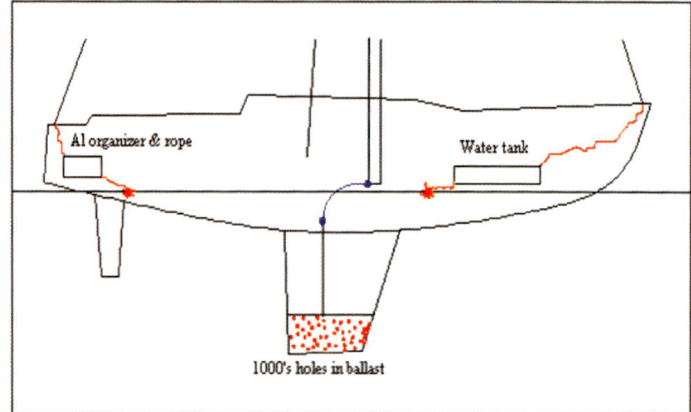

Galvanic Corrosion

When two different type metals are in contact and subject to a corrosive environment, the least noble metal will be sacrificed.
A corrosive environment would include submersion in salt water or even freshwater, or when subjected to seawater spray.

The process of accelerated corrosion begins because of an exchange of ions, electrons, and other atomic and subatomic particles at the point where these metals touch.
This exchange of particles at the junction of these metals causes an electrical difference of potential between the metals.
Mast bases are notorious locations for galvanic corrosion due to the propensity for differing metals to inhabit these spaces.

Testing Shore Power & Boat Grounds

Galvanic current flowing around the circuit will corrode the least noble metal between the two (or more) boats, in this case an aluminum sterndrive

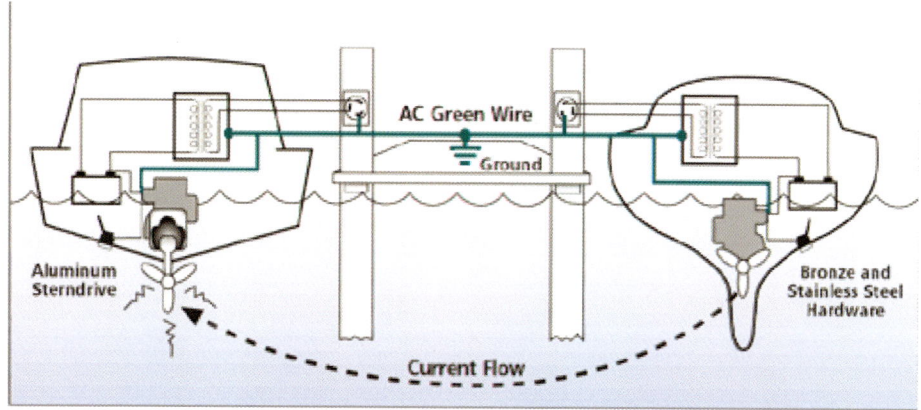

Electrolysis

People generally do not understand this term, using it as a catch-all to describe any kind of corrosion below the waterline.
Electrolysis is simply the result of stray current, and nothing else. Galvanism and electrolysis produce similar results, only they have different causes.
We would be better off using the term "stray current corrosion" because this identifies the cause.

Electrolytic Corrosion

Electrolytic corrosion is basically galvanic corrosion accelerated by the presence of an external source of current.
Comparatively speaking, seawater is a pretty good conductor of electricity.

So all the underwater exposed metal parts of your boat are electrically connected to all the underwater exposed metal parts of any other boats within a reasonable distance.
A mistake or fault in the wiring supplying power to your boat, the marina docks or any of the other boats nearby could cause a difference of potential between the submerged metal parts in your boat and the submerged metal parts connected to the faulty wiring.

Electrolytic Corrosion

Electrolytic corrosion is the beast that lurks beneath the surface. No matter the size of the marina, if boats sit in close enough proximity to one another, ions will be changing locations.

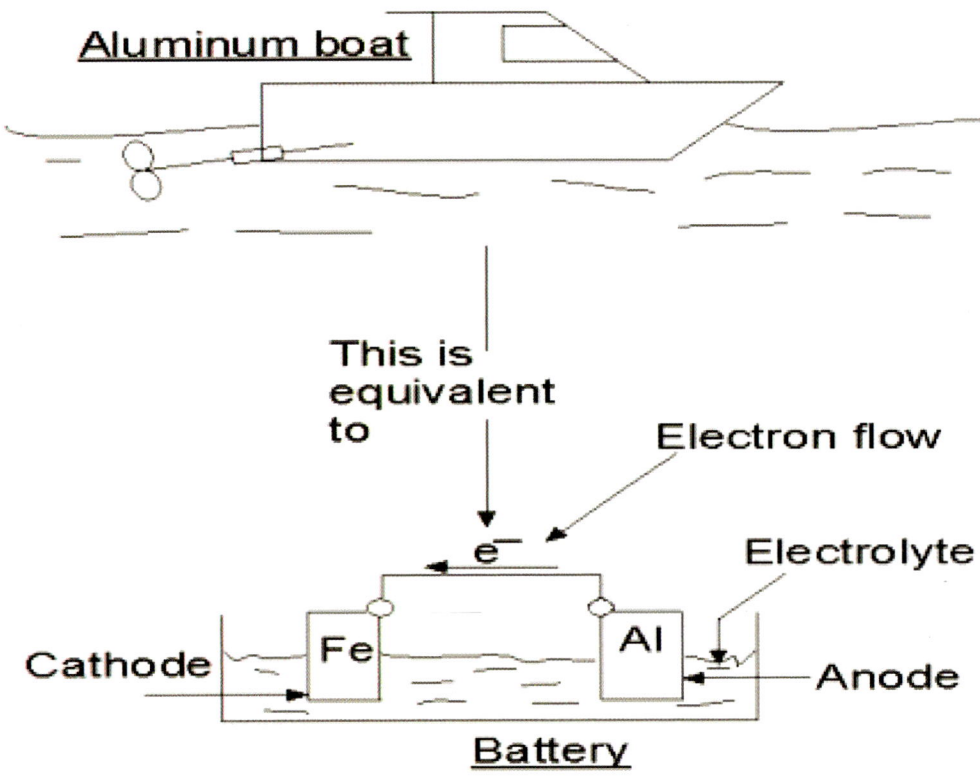

Boat Corrosion Issues Analysis

CHAPTER 11

AC Wiring

TOPICS

11.0 50 Hz or 60 Hz	136
11.1 AC Wiring	139
11.2 AC Wiring Color Code	141
11.3 Types of Marine Wire	143
11.4 Neutral Connection	145
11.6 Grounding Conductor	147
11.7 Types of Outlets	148

Video Episode 1: AC current- Frequency & wire color code

Scan this code to see the highlight video

In this episode we will learn about the fundamentals of Frequency, also we will study the AC wiring color code for American Power 60 Hz and European power 50 HZ and finally we will analyze the concept of efficiency in terms of output power for both types of currents.

Follow me

60 Hz Vs 50Hz

$$n = \frac{\#\text{ of Poles} \times RPM}{120}$$

American Power $60 Hz = n = \frac{4 \text{ Poles} \times RPM}{120} \Rightarrow RPM = 1800$

European Power $50 Hz = n = \frac{4 \text{ poles} \times RPM}{120} \Rightarrow RPM = 1500$

Four Poles Rotor

This is a typical marine generator with 4 Poles at the rotor. The rotor is bolted on the flywheel of the engine. Other words the rotor is spinning at the same RPM of the engine.

50 Hz vs 60 Hz / Master Class Part I

Why the Efficiency is less in 50Hz

Suppose that you have a generator of 20Kw, 240V, 80 Amp and 60Hz.
What happen with the output power if you convert the generator to 230 V, 80 A, 50 Hz.

$$\text{American Power 60Hz} \Rightarrow P = V \times I = 240_v \times 80_A = 19200 \text{ watts}$$

$$\text{European Power 50Hz} \Rightarrow P = V \times I = 230_v \times 80_A = 18400 \text{ watts}$$

The output Power is less in 50Hz

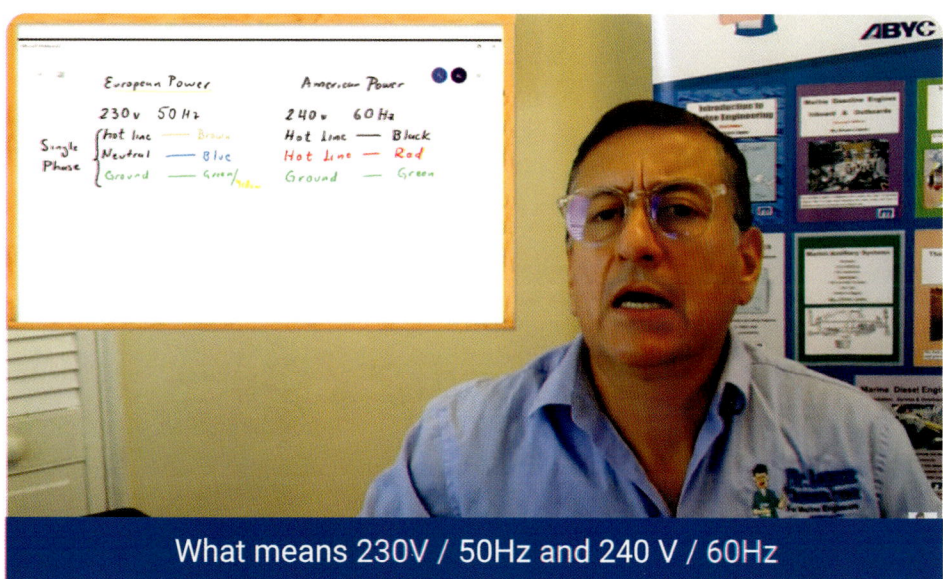

What means 230V / 50Hz and 240 V / 60Hz

230V-50Hz or 120/240V-60Hz

In short, 50 Hz vs 60 Hz is more likely similar to the discussion between Coca Cola and Pepsi Cola.
Thomas Edison (USA) calculated that 60 cycles per second or 60Hz was the most effective frequency.
Nikola Tesla (German) later compromised to reduce the voltage to 110 v 50Hz for safety reasons.
Unfortunately, 50Hz AC has greater losses and is not as efficient as 60HZ. Due to the slower speed 50Hz electrical generators are 20% less effective than 60Hz generators.

Chapter 11: AC Wiring

Efficiency in 60Hz vs 60Hz

Electrical transmission at 50Hz is about 10-15% less efficient.
50Hz transformers require larger windings and 50Hz electric motors are less efficient than those meant to run at 60Hz.
They are more costly to make to handle the electrical losses and the extra heat generated at the lower frequency.
Europe stayed at 110V AC until the 1950s, just after World War II. They then switched over to 230V (1 Phase) for better efficiency in electrical transmission

Wire Color Code in 50 Hz vs 60Hz

- American Power Single Phase
  120 V - 60 Hz

- European Power Single Phase
  230 V - 50Hz

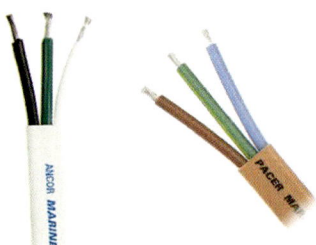
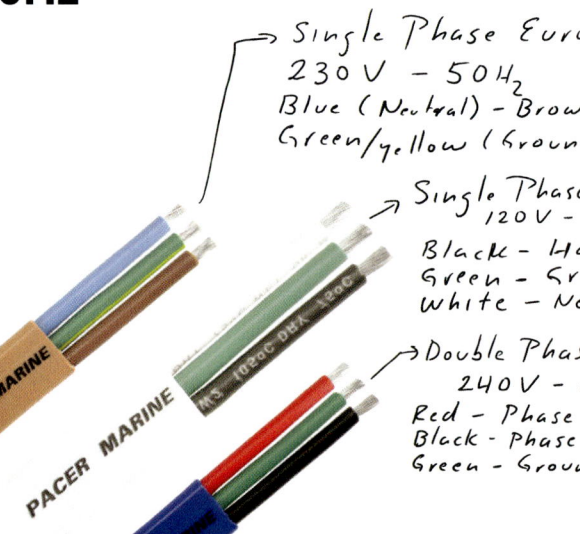

Single Phase European
230 V - 50 Hz
Blue (Neutral) - Brown (Hot)
Green/yellow (Ground)

Single Phase American
120 V - 60 Hz
Black - Hot
Green - Ground
White - Neutral

Double Phase American
240 V - 60 Hz
Red - Phase 1
Black - Phase 2
Green - Ground

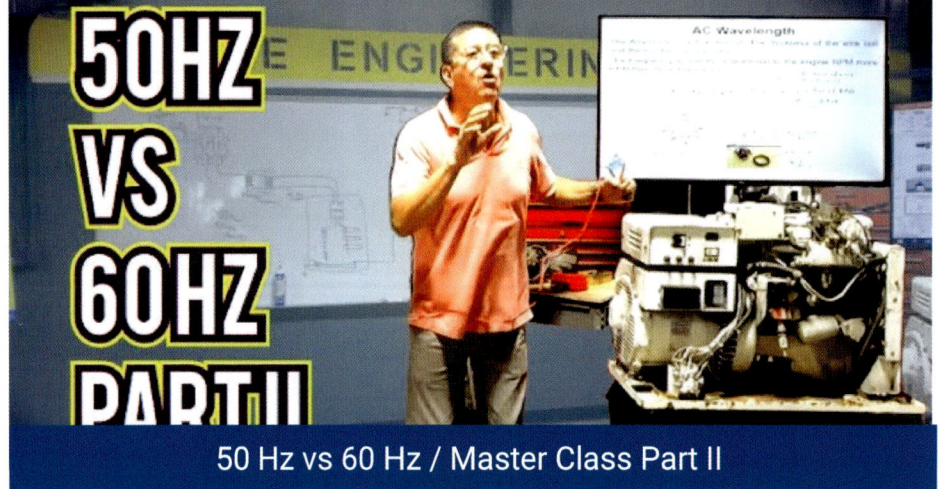

50 Hz vs 60 Hz / Master Class Part II

230V-50Hz or 120V-60Hz

What happens if you connect a 60 Hertz motor to a 50 Hertz mains?
1. The motor is 17% slower
2. The internal current goes up by 17%
3. The power (watt) goes down by 17%
4. The mechanical cooling is less, because of 17% fewer turns.

The result is a higher current than the manufacturer's design and the insulation of the electrical wiring deteriorates much quicker, which results in a burn-out.

U.S. stays at 120V, 60Hz

The United States also considered converting to 240V for home use but felt it would be too costly, due to all the 120V electrical appliances people had.

A compromise was made in the U.S. in that 240 V would come into the house where it would be split to 120V to power most appliances.

Certain household appliances such as the electric stove and electric clothes dryer would be powered at 240V.

Examples 50Hz/60Hz

A hairdryer has a heating element and a fan-motor, the heating element is not a problem, but the fan-motor is made for 60 Hertz, so can burn-out

A mains adaptor for a battery charger, for a laptop, or for a cellular phone has a transformer in it and burns-out if it is only made for 60 Hertz

Would 230 volts 50 Hz appliances work on 240 volts 60 Hz?
In general the answer has to be: "Not if it has been designed and wired to run only on the 50 Hz mains frequency system that is used in Europe and elsewhere.

AC Wiring (120V / 60Hz)

In the United States, commercial and domestic wiring follows a color code established by the National Electrical Manufacturers Association (NEMA).

Three wires are used for standard 120 volt circuits. A **black** color is used for the "Hot Line," a white wire is the "Neutral," and a green wire is used for earth ground

The main shore cable also has a **black**, white, and green wire. The **black** and white wires go to the vessel's main circuit breaker, while the green wire goes to the vessel's ground buss.

Marine Grade Wire

Marine grade wire is available as either boat cable or in spools of stranded insulated wire.
Boat cable consists of either two or three insulated stranded wires within a white vinyl outer wrap, and is clearly marked as marine boat cable.

Marine Duplex Wire

These can help simplify wire installation by reducing the number of separate wires run throughout a boat.
Different color codes are used depending on the application. AC wiring uses **black** & white. DC uses **black** & **red** or **red** & **yellow**. Red & yellow is the safety color to eliminate the risk of connecting black AC with black DC. Typically used for DC main power take offs, DC electrical switches & controls, AC wiring.

Marine Triplex Wire

Another common multi-conductor boat cable, triplex wire an also reduce time spent in installation. The usual color code for triplex wire is **black**, white & **green** (designed for AC as hot, neutral & ground). Typically used for AC wiring.

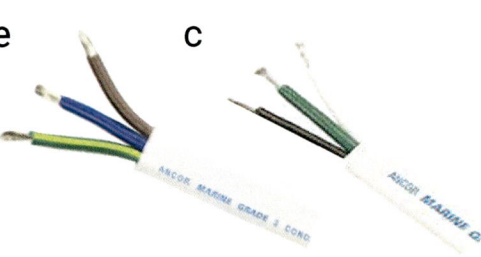

Colors for AC Power

European 50 Hz Brown, Blue & Green/Yellow

Black or brown to AC "HOT"
White or blue to AC "NEUTRAL"
Green or green/yellow to AC "GROUND"

Colors for AC Power

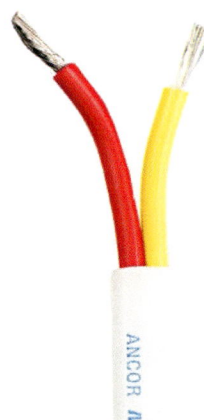

Yellow wire is recommended as DC negative because black is the standard color for AC hot. There have been many cases of people working on their DC systems who have inadvertently cut the live AC wire.

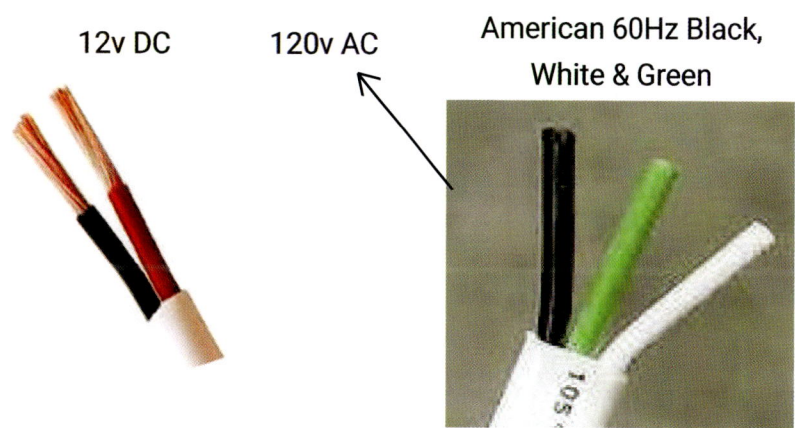

American Wire Gauge

Sizes vary according to the application, with lower numbers indicating the ability to carry heavier current loads.
For example, #4 AWG is commonly used for high amperage battery connections and #14 AWG for "house" wiring.
Wire size is as important as everything else. Always use the proper size wire for the electrical load it will be carrying.
For circuits which are relatively short use a wire gauge one size larger than the size wire the accessory you are connecting provides.

AWG Wire gauge Vs Amps

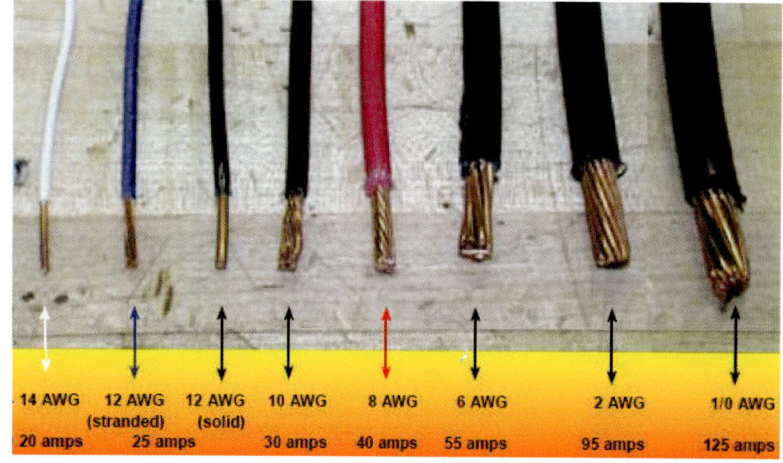

Chapter 11: AC Wiring

Wiring a Single Phase Boat

AC Wiring (120V / 60Hz)

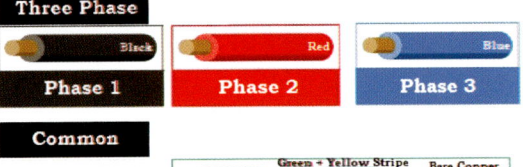

Four wires are used for standard (American) 240V/60Hz volt circuits. A **Black** color is used for the 120 volt "Hot Line(1)," a **Red** color is used for the 120 Hot Line(2), a White wire is the "Neutral," and a Green wire is used for earth ground

The main shore cable also has a **Black**, **Red**, White and Green wire. The **Black**, **Red** and White wires go to the vessel's main circuit breaker, while the green wire goes to the vessel's ground buss.

Function	label	Color, common	Color, alternative
Protective ground	PG	bare, green, or green-yellow	green
Neutral	N	white	grey
Line, single phase	L	**black** or **red** (2nd hot)	
Line, 3-phase	L1	**black**	brown
Line, 3-phase	L2	**red**	orange
Line, 3-phase	L3	**blue**	yellow

Types of Marine Wire

The smaller AWG (American Wire Gauge) sizes (16 to 8 AWG) of single conductor wire are called primary wires.
These usually make up the circuit wiring of a boat's auxiliary systems to the main battery.
Typically used for DC electrical switches & controls.
Large (6 to 4/0 AWG), single conductor wire is referred to as battery cable.
The cable usually has a high tin plated copper strand count. The tin plated copper helps minimize corrosion and extend the life of the cable. Typically used for battery installation and DC main power take offs.

Voltage vs Wire Gauge

110 -120-volt, 60 hertz, properly grounded branch circuit protected by a 15/20-amp circuit breaker or fuse. Must be properly grounded and polarized. #14-gauge house wire minimum for 15-amp protection, #12 gauge is a must for 20-amp protection
220 - 240-volt, 60 hertz, properly grounded circuit with 40-amp breaker or fuse protection with #8-gauge wire. With a 50-amp breaker, # 6-gauge wire is required.

Contact between Conductors

Whether or not this presents a shock hazard depends on *which* wire accidentally touches:
If the "hot" wire contacts the case, it places the user of the toaster in danger.

On the other hand, if the neutral wire contacts the case, there is no danger of shock.

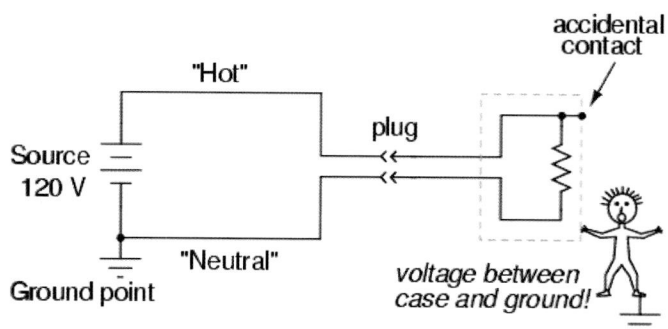

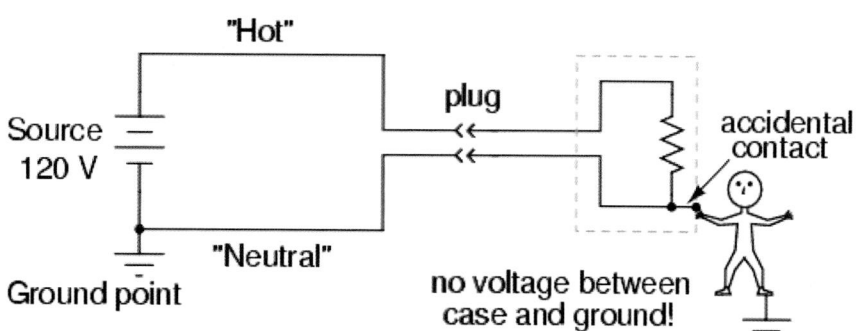

AC Wiring (230V / 50Hz)

Three wires are used for standard (European) 230 volt circuits. A **Brown** color is used for the 110 volt "Hot Line," a **Blue** wire is the "Neutral," and a **Green** wire is used for earth ground.

The main shore cable also has a **Brown**, **Blue**, and **Green** wire. The **Brown** and **Blue** wires go to the vessel's main circuit breaker, while the **green** wire goes to the vessel's ground buss.

European Color Code

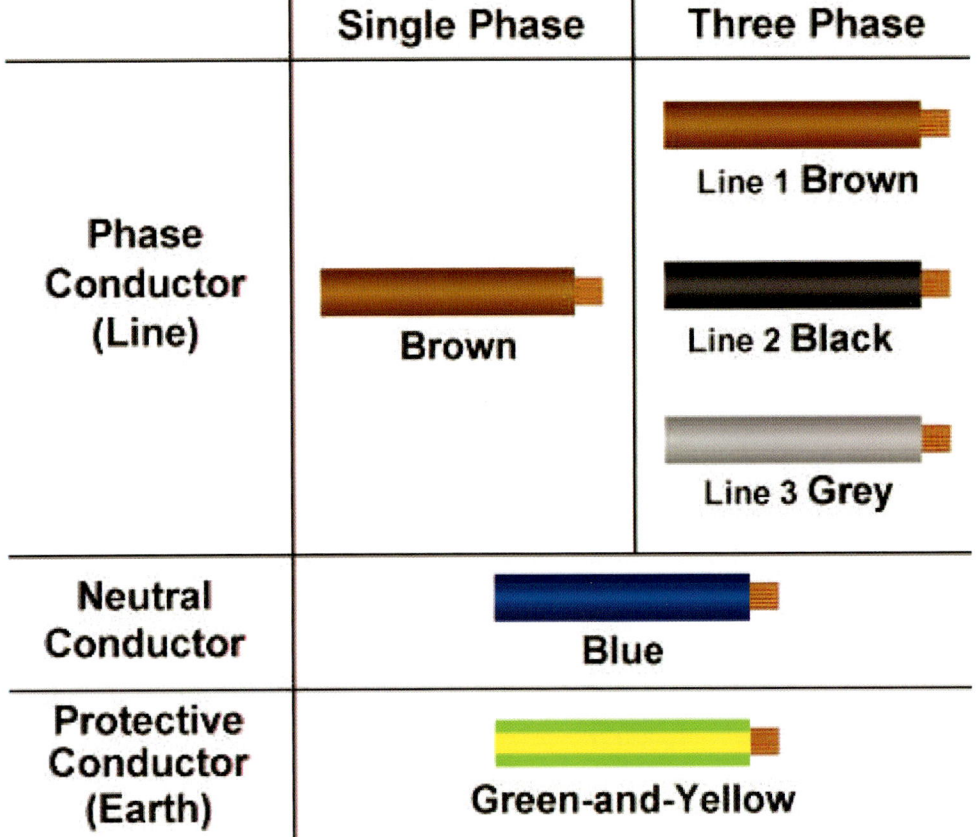

AC Wiring (230V / 50Hz)

Three wires are used for standard (European) 230V/50Hz volt circuits. A **Brown** color is used for the 230 volt "Hot Line," a **Blue** wire is the "Neutral," and a **Green**/**Yellow** wire is used for earth ground.

The main shore cable also has a **Brown**, **Blue** and **Green** wire. The Brown and **Blue** wires go to the vessel's main circuit breaker, while the **green** wire goes to the vessel's ground buss.

How the Neutral wire is connected

In a 230 Volts 50 Hz appliance there are 3 wires, a "**Ground**" conductor, one single 230 volt "hot" wire and a **Neutral** wire connected as a return to the source.

There is a 230 Volt voltage difference between the "**Neutral**" and the "**Hot**" conductor in the 50 Hz system and only a **120 Volt** voltage difference between the "Neutral" and each of the "Hot" conductors in the 60 Hz system. negative).

In the main breaker box, at the point where the 60 Hz "Neutral" gets connected to the "Ground", this difference will cause serious problems!

That is why an appliance designed to be connected onto the 50 Hz system cannot be used safely on the 60 Hz system without first having a proper technical inspection.

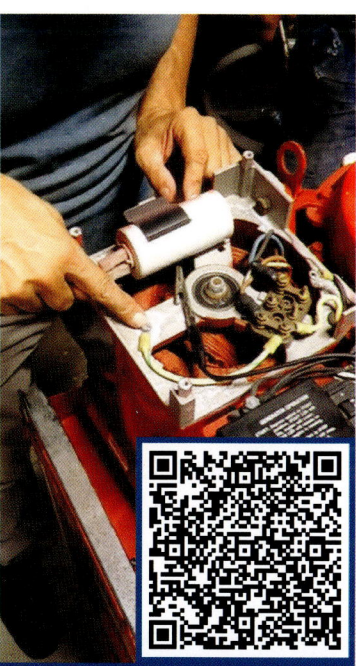

Neutral and Ground Together at the Generator

Some Exceptions

Some small "double-insulated" 50 Hz appliances, such as electric shavers, etc., have been designed to run **safely** on different supply voltages and frequencies. If that is so, it would be stated on their rating plates.

In many cases, where the power needed is low, such as (say) less than 30 Watts, a cheap and simple "International Travel Socket Adapter" is all that is needed to make such an appliance plug-in and work. Many international **airports** have shops selling such adapters.

Neutral and Ground

AC current must alternate between two points making a circuit.
Coming from the power source is the "hot" wire, which normally has black insulation, and returning is the neutral or "grounded" conductor, which is white. **"Neutral" carries the same current as the hot wire.**
As long as the current remains in this closed circuit there is no danger, but if it should escape (a "fault" or "short circuit"), it will attempt to go directly to ground.
In a boatyard, where workers may be standing or crawling on wet ground, there is a potential for electrocution. When a boat is floating, the water is the ground and any metal that has an electrical path to it, including the hull of a metal boat or the engine of a glass or wood boat via the shaft, becomes a path to the ground. Touching any of these items and a **hot wire** at the same time can send current through the body.

Neutral & Ground Tie-In

With 120 V systems, the green and white get tied together at any AC power source, but only at a source of AC power
Next consider potential power sources used for boats:

- Generators
- Inverters
- Shore power delivered from the dock

According with ABYC Neutral and ground shall be connected together only at the source of power. After that both conductors should be isolated.

Inverters are only tied together when the inverter is operating in AC production mode.

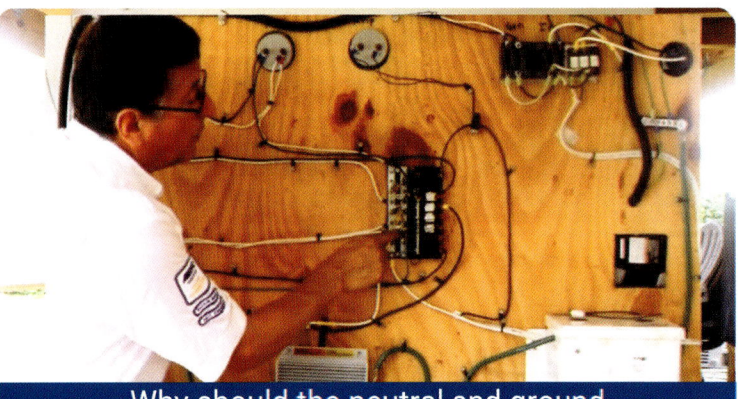

Why should the neutral and ground be connected at the power source?

The shore power neutral is grounded through the shore power cable and shall not be grounded on board the boat.

Grounding conductor

When fault current is conducted to ground, the immediate result should trip a breaker or blow a fuse, effectively shutting off the power.
The green grounding conductor must be sized so it's capable of safely carrying fault current without overheating. This conductor could be thought of as insurance.

Grounding Path

Types of Outlets

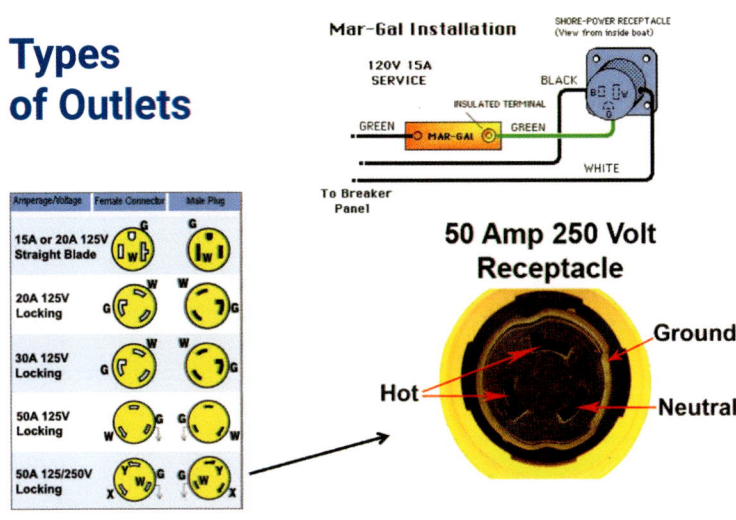

Contrary to popular belief, electricity takes all paths back to its source, not just the path of least resistance.
This presents a clear hazard to swimmers, and those aboard may also be at risk, depending upon the nature of the fault
Therefore, it is vitally important that the **connection between the AC safety grounding wire and the neutral wire occur only at the power source !!**

Neutral and Ground Wiring

All the AC circuits have a **green** third wire, which is a "grounding" wire.
It is connected to the third prong of the common three-prong plugs; it parallels the white wire and it connects to neutral at the power source.

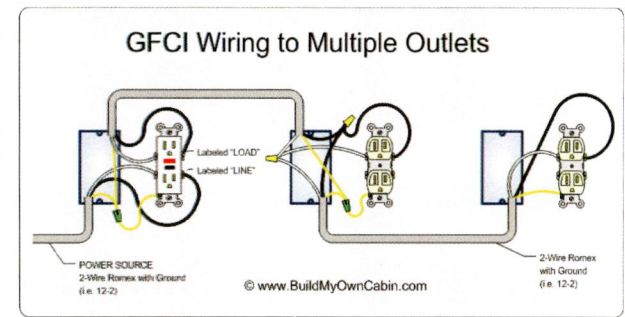

AC Outlet Wiring

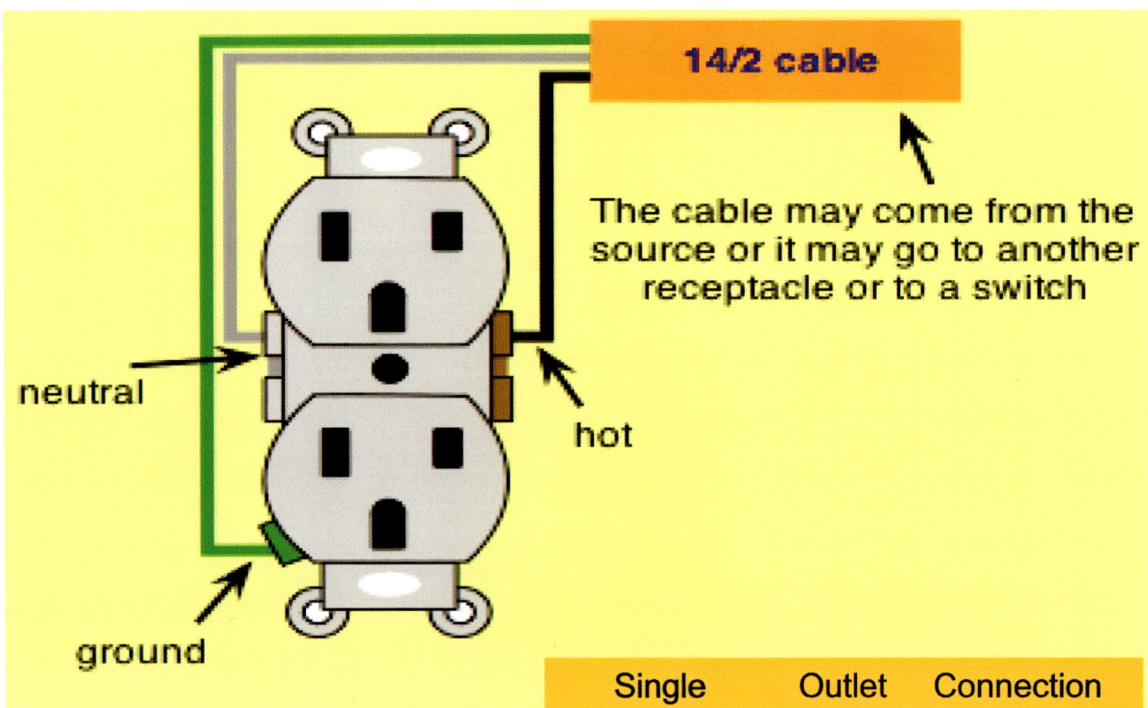

Two or More Outlets in the Same Circuit

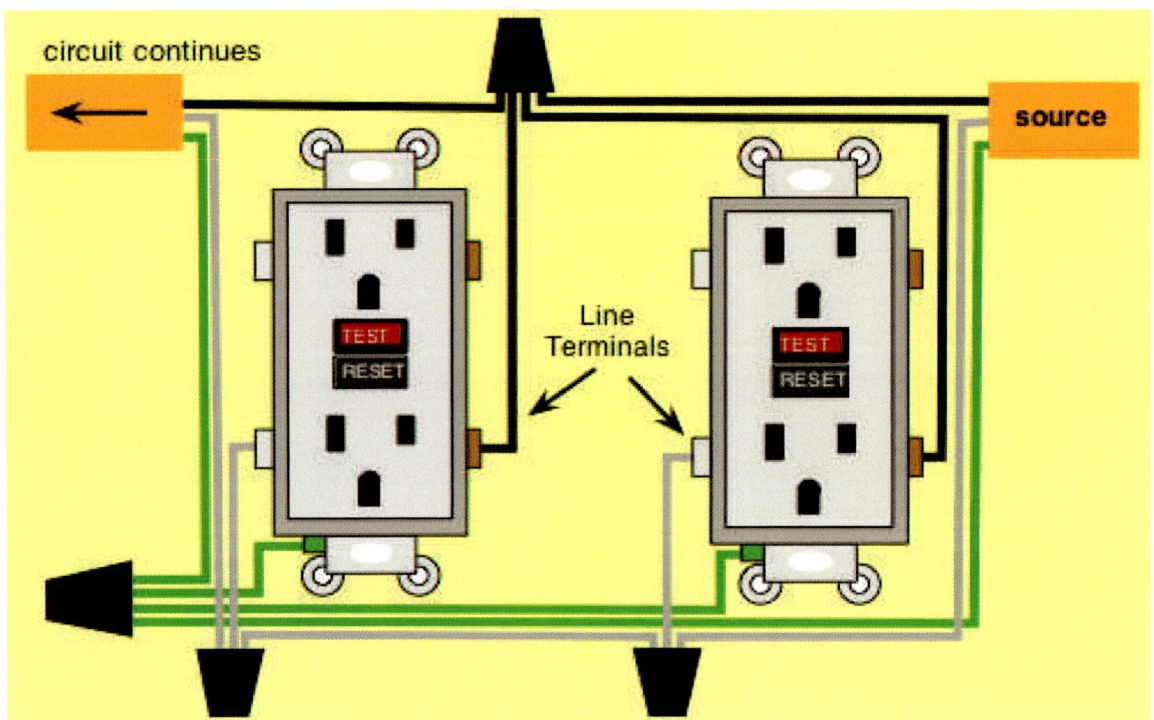

CHAPTER 12

AC Panel board

TOPICS

12.0 Ship / Shore Switch	150
12.1 Triple breaker Switch	151
12.3 Double Phase Wiring	152
12.4 Double Phase Open Neutral	153
12.5 Double Shore cord Inlet	153
12.7 Two shore Cords / Two Generators	154
12.8 Reverse Polarity Indicator	155
12.9 AC Wiring Warning	159
12.10 E.L.C.I	161
12.11 Faulty Ground	161
12.12 Shore power Cord	162
12.13 Shore power Connectors	163
12.14 Shore power Pedestals Wiring	166

Scan this code to see the highlight video

Video Episode 1
Wiring an AC Panel / Reverse Polarity Diagnosis

This video is considered one the best sellers in marine electrical installations. In the video you will learn about the wiring procedure of an AC panel double phase including: voltmeter, amp-meter, frequency meter, isolator transformer and galvanic isolator.

Follow me

Chapter 12: AC Panel Board

Ship/Shore Switch

A proper system should include a "ship/shore" switch so the vessel can be powered from either shore power or generator power.

Connect both the black "Hi side" and the white "Low side" wires to a set of "input" terminals on the switch.

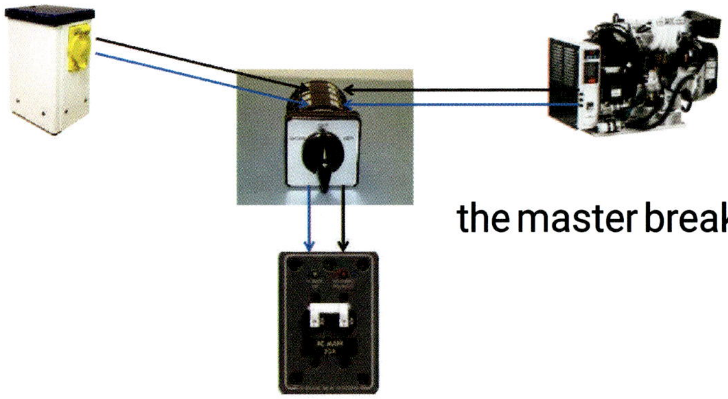

Disconnect the shore power cable from the master circuit breaker, and reconnect it to the other set of "input" terminals on the switch. The "output" set of terminals is then connected to the master breaker.

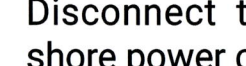

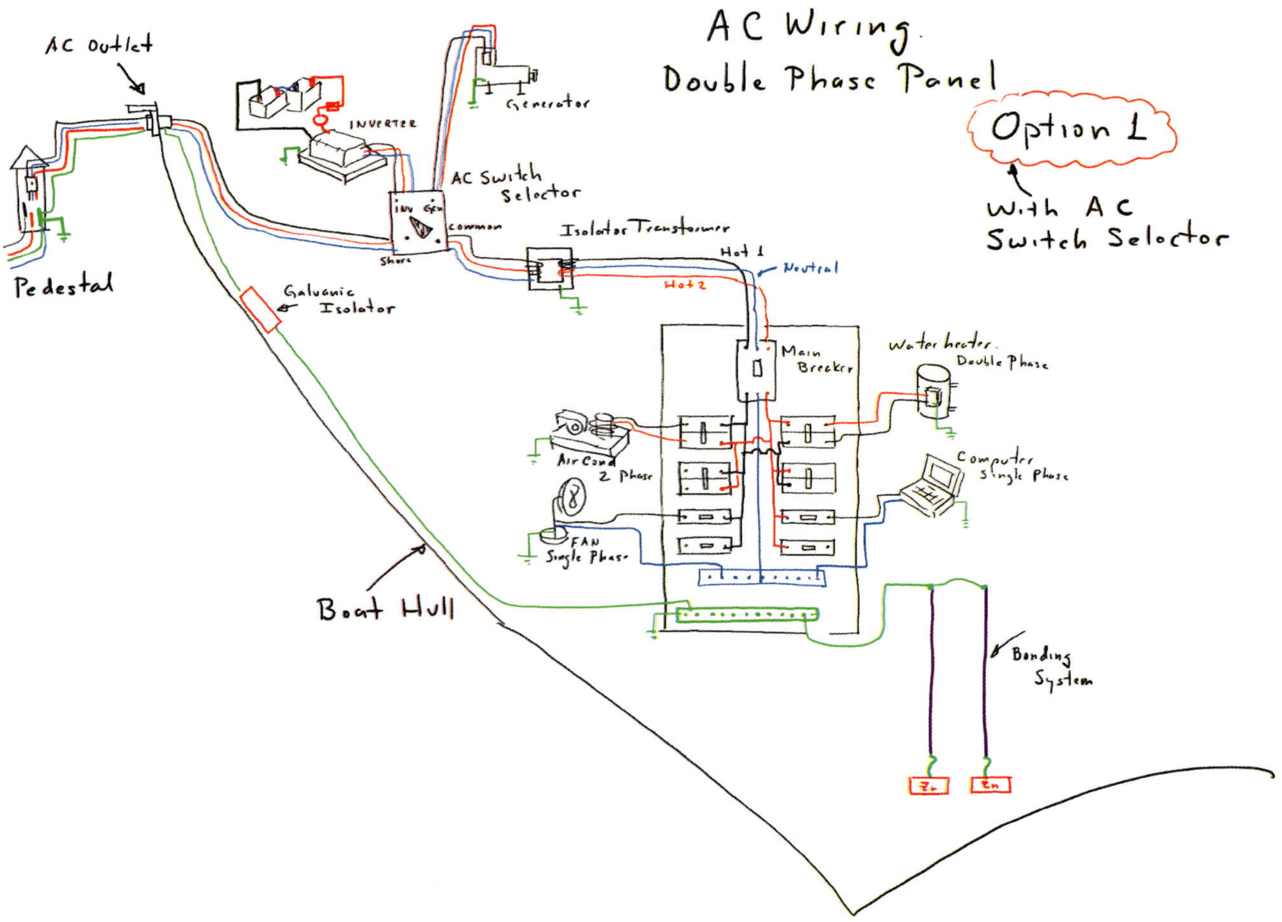

Triple Breaker Switch Selector

Some boats come with a triple breaker configuration instead of a switch selector.

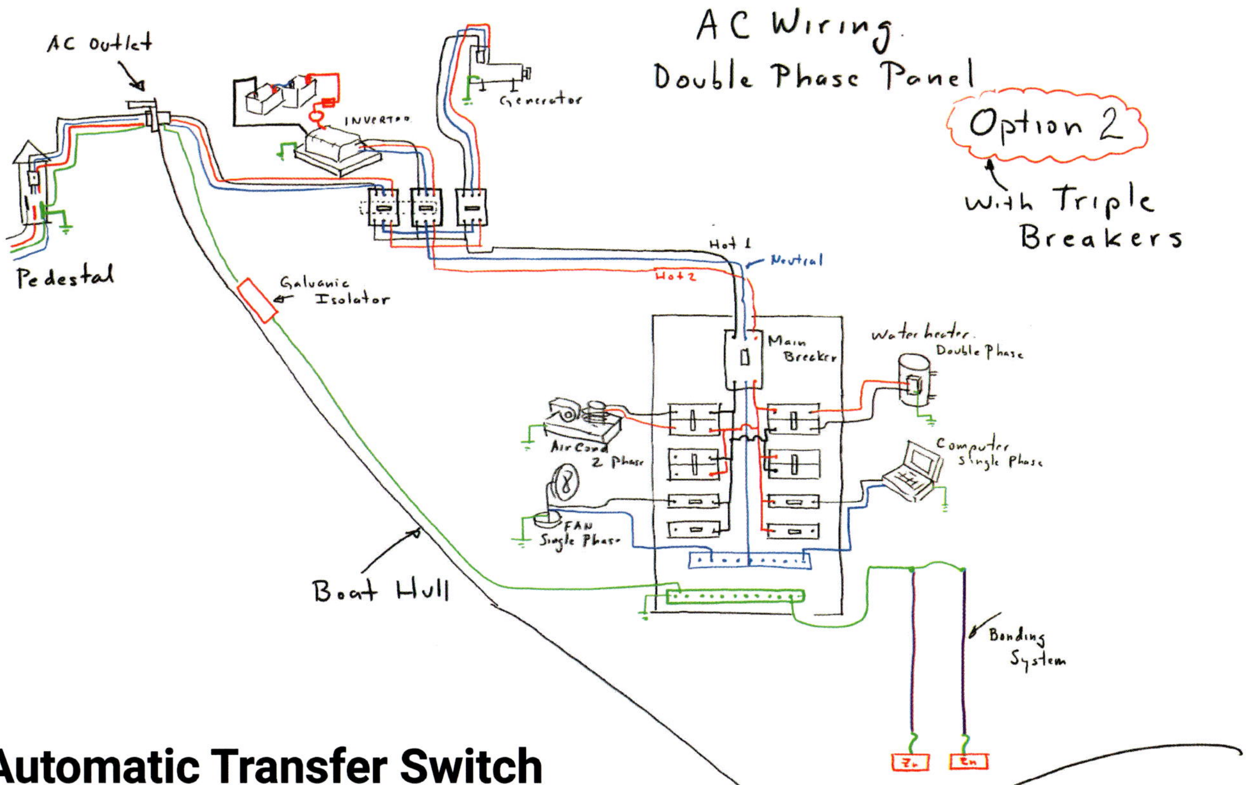

The Automatic Transfer Switch

Auto-transfer switches ensure that only one input power source is connected to the loads (AC Panel) at any time, but that one is always connected. They are often used with Inverters. In some cases with generators those cases, a time delay relays are quired to apply for the transfer once the generator be stable.

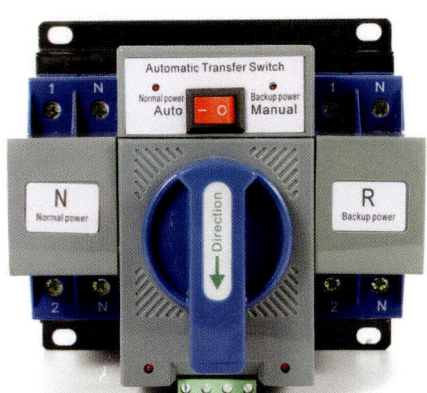

The Automatic transfer switches are often used with Inverters, which do not have their own internal transfer relays, to select between Shore Power and Inverter or Generator power. This type of system is more accurate every day however in comparison with the manual switches is too expensive. It is popular in mega-yachts.
This type of transfer switch can operate in automatic mode or manual mode.
The manual mode is commonly used in cases where the load stabilization time of generators is out of range.

Chapter 12: AC Panel Board

Two Phase Wiring (240 V 60Hz)

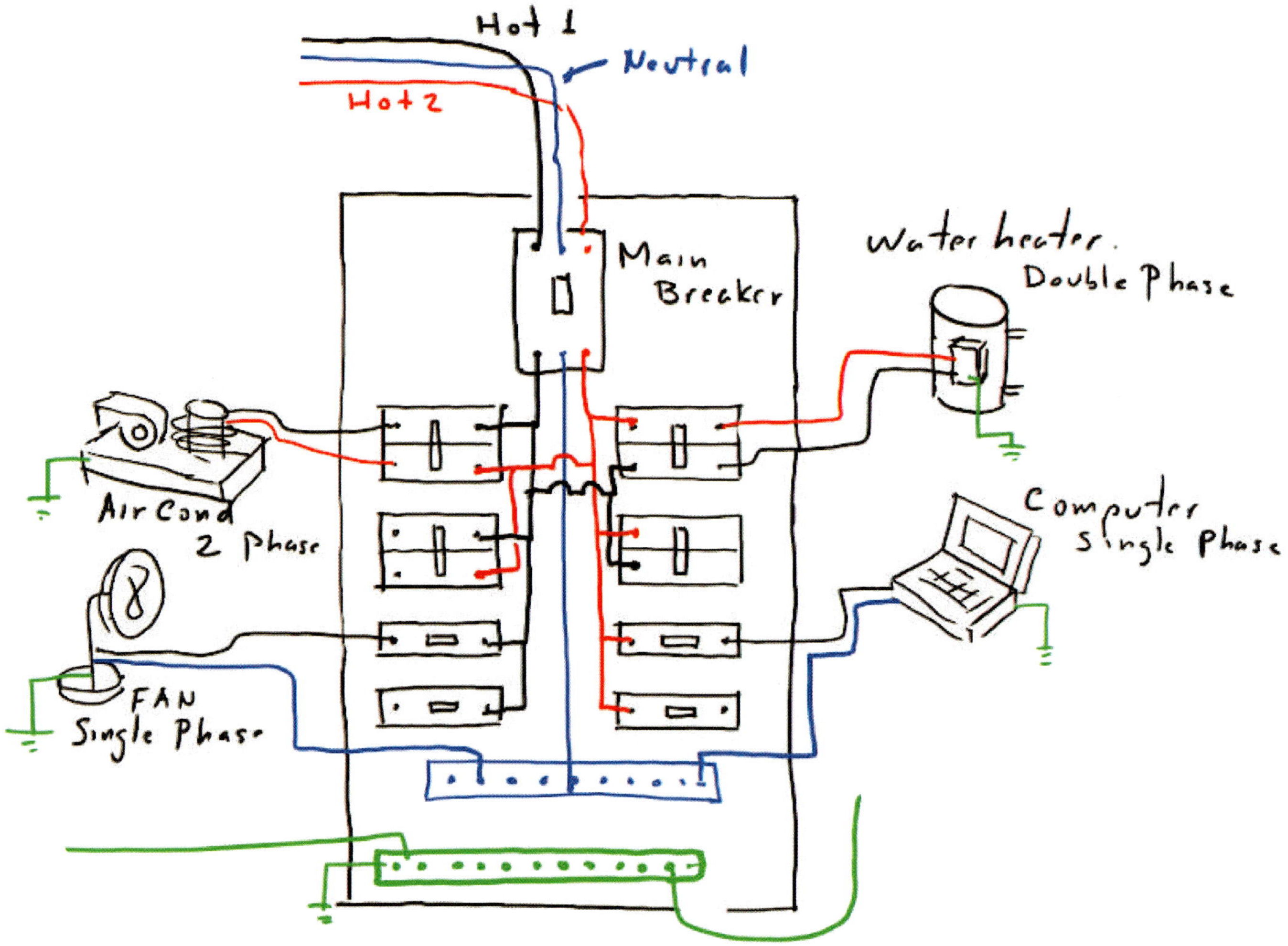

How to Wire a Boat Double Phase Panel

Chapter 12: AC Panel Board

2 Phase Open Neutral

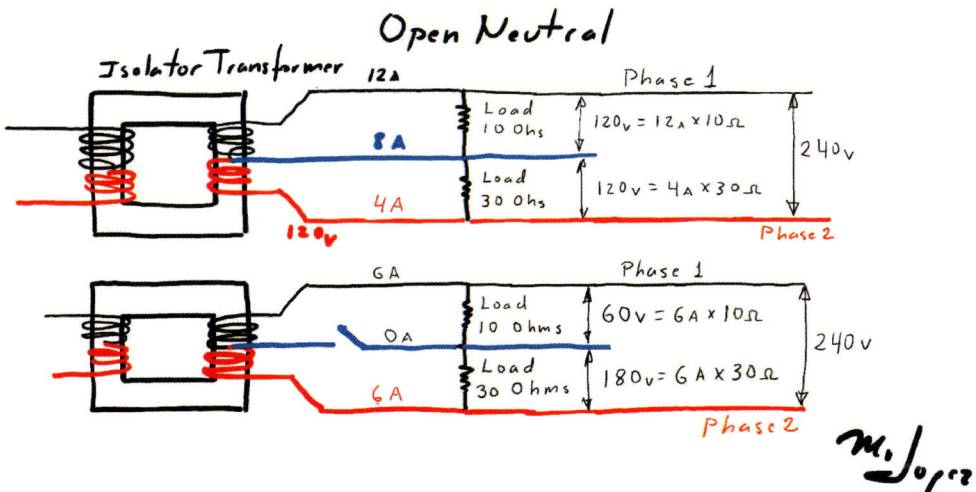

Double Shore Cord inlet

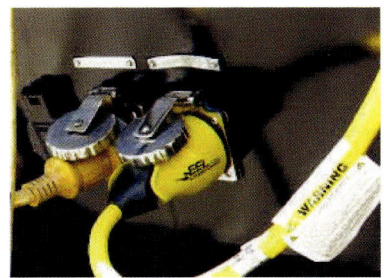

If one cord is connected to shore power, one of the terminal lugs under the cap of the other inlet will be live, representing a shock hazard.

More than one Shore Power Inlet

The shore power neutrals shall not be connected together on board the boat.
The current carrying conductors must be isolated either by completely separating, or via a break-before-make type of transfer switching device.

In dual shore cord using a single isolator, the current rating of the galvanic isolator shall not be less than the sum of the ratings of the main breakers.

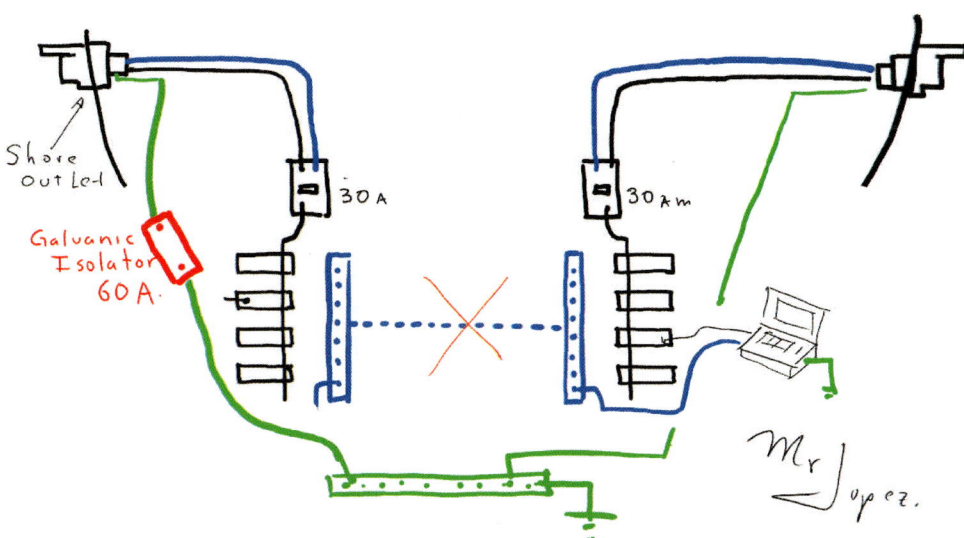

Chapter 12: AC Panel Board

Y to Divide the Capacity (Amp)

For power boats I recommend having a short mast with a lightning rod attached. If the mast is metal use 4awg wire at the base of the mast to connect through to our grounding source.

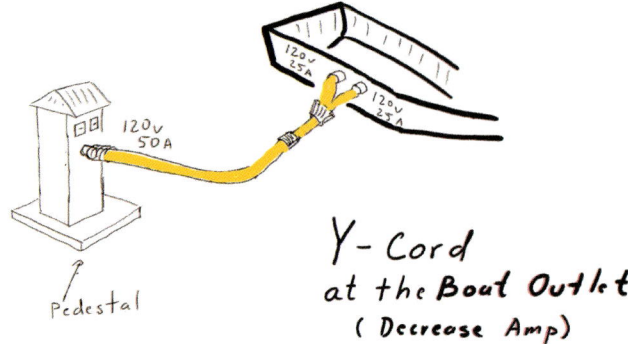

Y-Cord at the Boat Outlet (Decrease Amp)

Y Cord Increase (Amp)

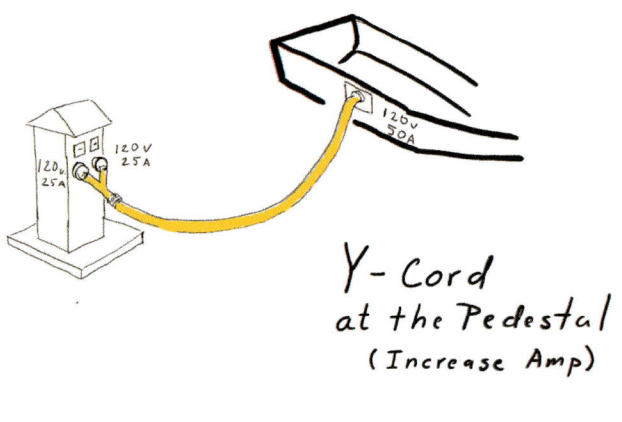

Y-Cord at the Pedestal (Increase Amp)

How a Dock Pedestal works

2 Shore Cords / 2 Generators

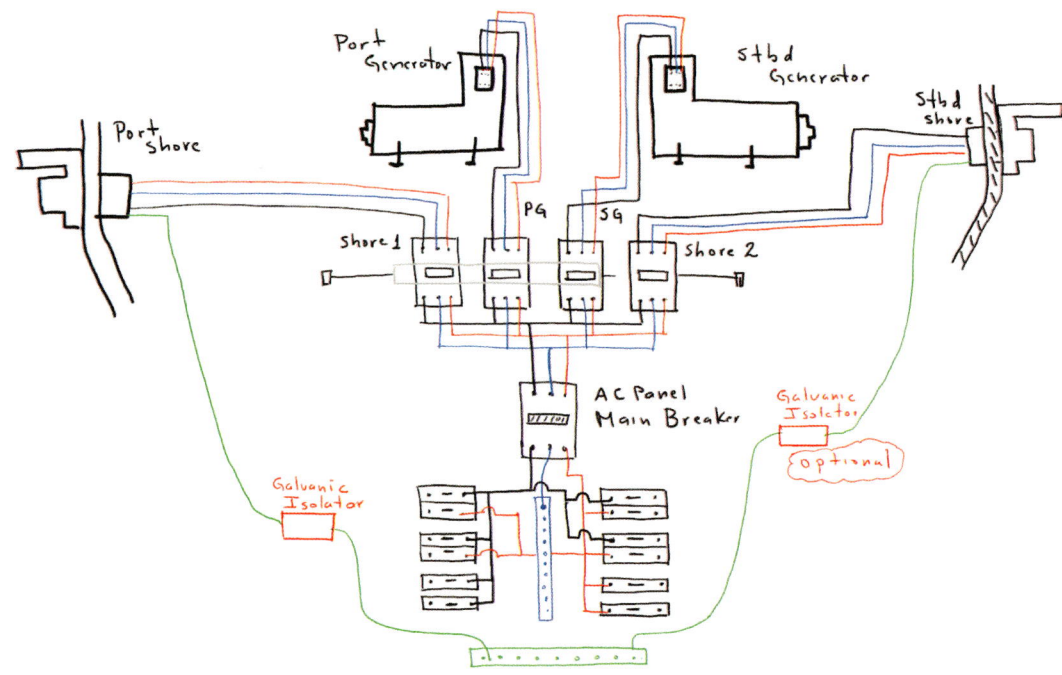

Reverse Polarity

Solving a Reverse Polarity Issue in a Sailboat

Reverse Polarity Indicators (RPI)

The job of the RPI is to determine if there is voltage potential between Safety Ground and Neutral.

One further ABYC requirement for RPI's is that they contain a minimum 25,000 Ohm Resistor
the Safety Green wire is connected to devices aboard the boat that may contain stray currents, the link created by an RPI could provide a path for stray currents via the grounded Neutral wire.

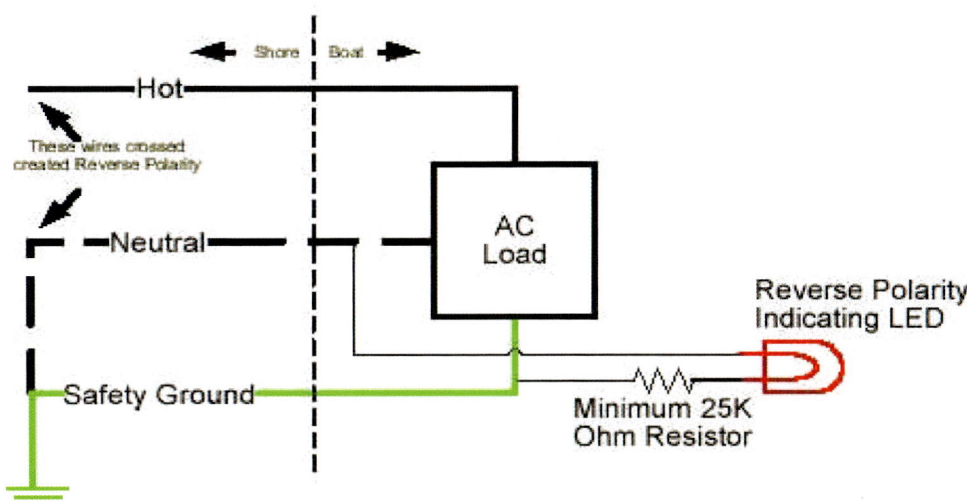

Chapter 12: AC Panel Board

Reverse Polarity Indicator

Perhaps the easiest way to visualize the working of an RPI is to first visualize the relationship of the Safety Ground and Neutral wires. The job of the RPI is to determine if there is voltage potential between Safety Ground and Neutral.

How to Connect the Reverse Polarity Indicator

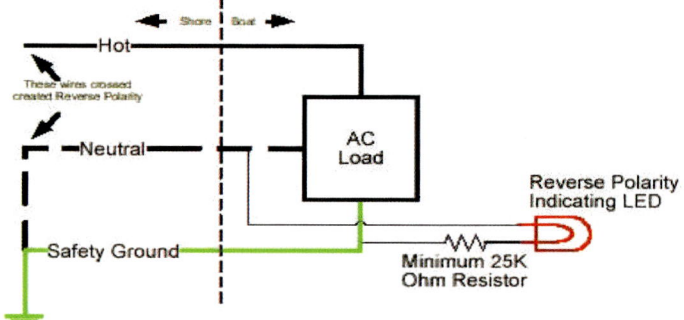

Polarity Indicators

Vessels equipped with polarity indicators can monitor the local dock power for correct polarity.

Many older dock power outlets do not follow the **NEMA** wiring code, and can be hazardous to you and your vessel. Some foreign built vessels do not follow the **NEMA** code either.

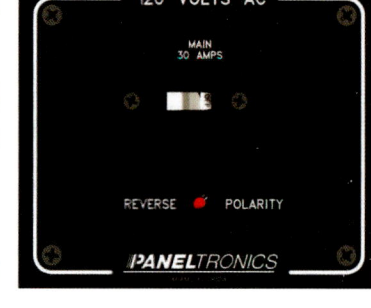

If the source of external 115-VAC power is incorrectly wired and has reversed the LINE and NEUTRAL conductors, it will present a voltage of 115-VAC between the NEUTRAL and GROUND conductors on the vessel.

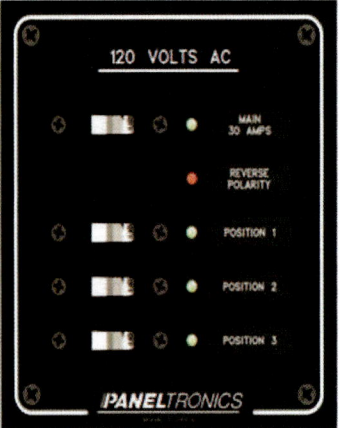

Reverse Polarity at the Power Cord

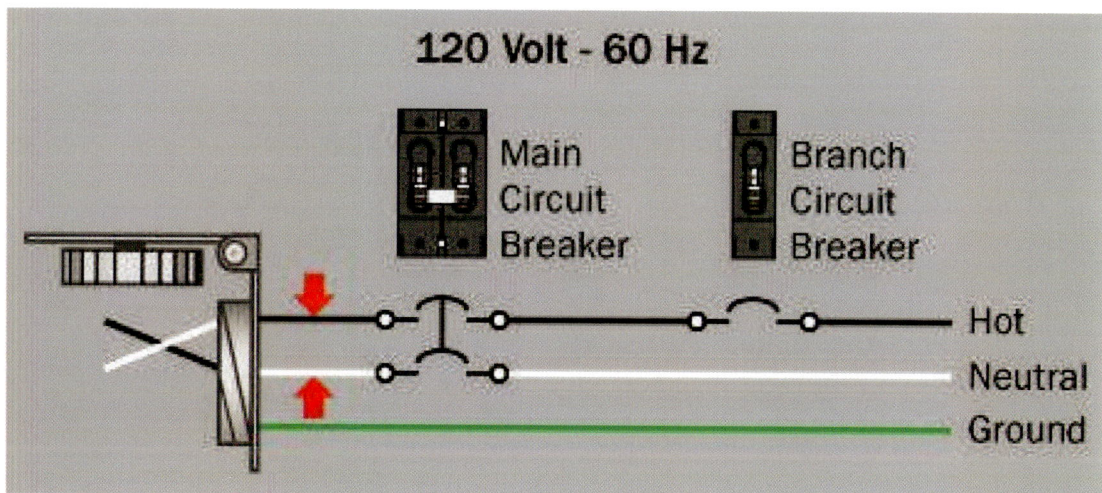

Polarity Indicator

There are two situations that eliminate the need for a polarity indicator on a 120 volt AC electrical system. One is if the boat is equipped with an isolation or polarization transformer.

These are large and expensive pieces of equipment that are generally only found on larger/medium yachts. The other exception is a boat that is equipped with double pole circuit breakers on every branch circuit. This type of breaker is designed to connect or disconnect both the hot and neutral wires for a given circuit simultaneously.

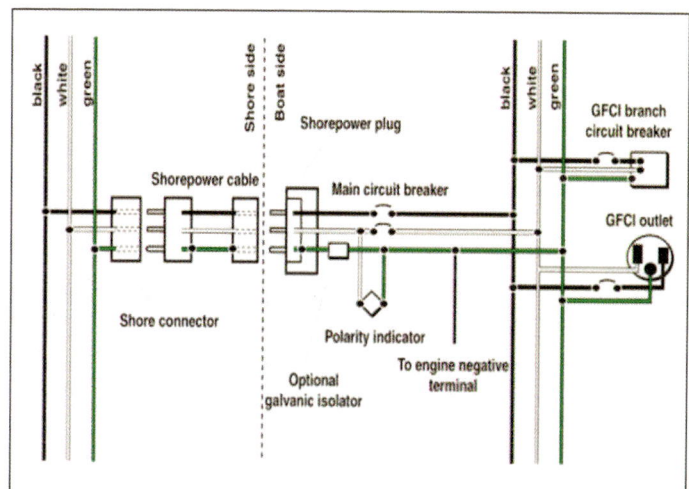

Some larger boats are equipped with a 240 volt AC shore power interface. Polarity indicators are not required on these systems because if the problem exists it will be abundantly obvious as things begin to blow when the connection is made.

The proper way to connect your boat to shore power is to turn off the main breaker on the boat and the breaker at the shore power pedestal. Then make the connections with the shore power cord. Once the connections have been made turn on the power at the pedestal/dock and then check the polarity indicator.

Chapter 12: AC Panel Board

What happens if polarity is reversed ?

If the polarity is reversed, then the neutral (the grounded conductor) is now powered by the hot (ungrounded conductor) line. All appliances and loads on the boat will still work since they don't care from which direction the alternating power comes from.

When a ground fault occurs, the current rises substantially on the neutral line, If there is no breaker in the neutral line (single pole), the conductor will heat up, cook, and could eventually cause a fire.

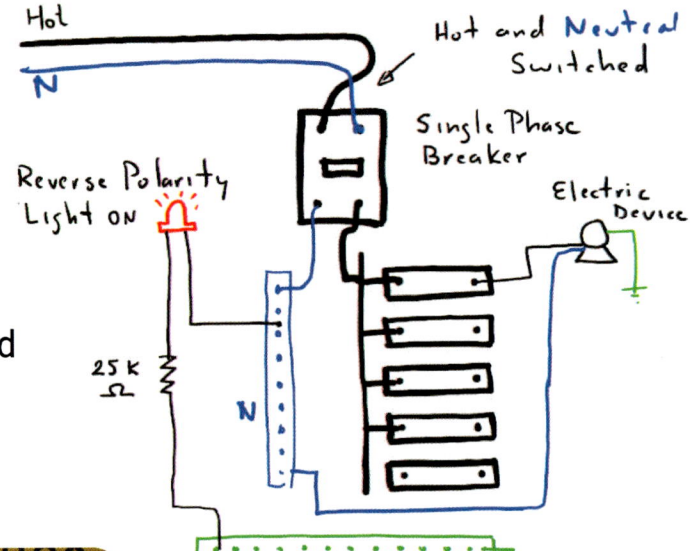

Single Phase Wiring and Reverse Polarity

What happens if polarity Ok?

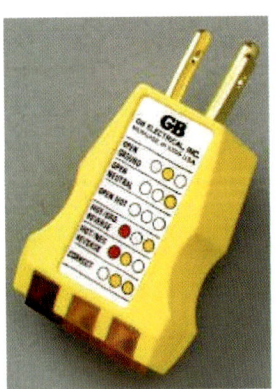

With proper polarity If a fault occurs in one of the loads (like a ground fault/short circuit), the current will increase substantially on the hot line which should trigger the branch supply circuit breaker for the load to trip.

That's the main purpose for having the ground system installed (to ensure there is a path for fault currents to go so that the current will become high enough to trip the breaker, at the same time keeping the voltage down on the exterior of the faulted component).

Chapter 12: AC Panel Board

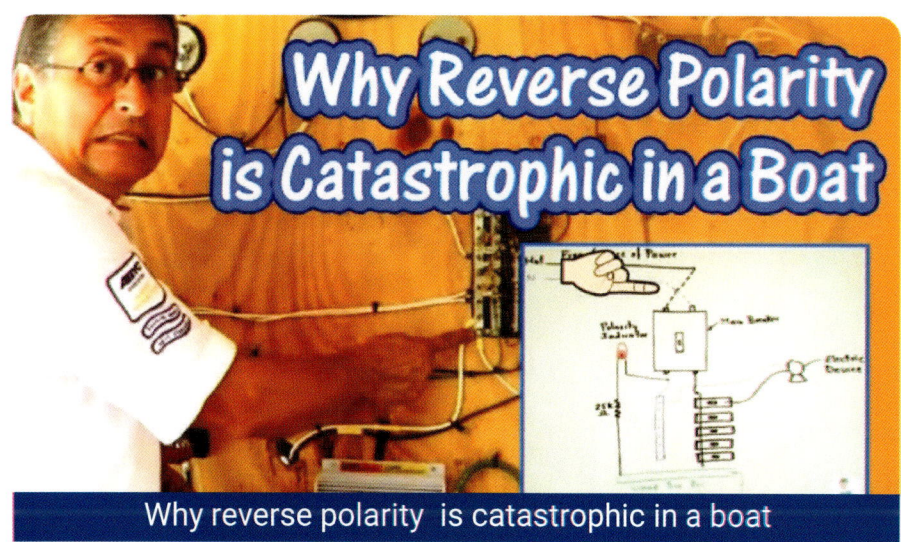

Why reverse polarity is catastrophic in a boat

AC Wiring (Warning)

The marine AC system is potentially more dangerous because the boat and the people who work on it are surrounded by water. A person who becomes part of the pathway between a hot wire and the sea can experience severe shock.

Remember, what makes the heart tick is a faint electrical impulse generated within the muscle itself. It takes only a very small amount of current through the chest to disrupt the heart rhythm, causing fatal fluttering of the heart muscle called fibrillation.

A critical factor is where the current passes through the body. Touching hot and neutral leads with one hand can give you a jolt and maybe even a burn, but won't kill you.

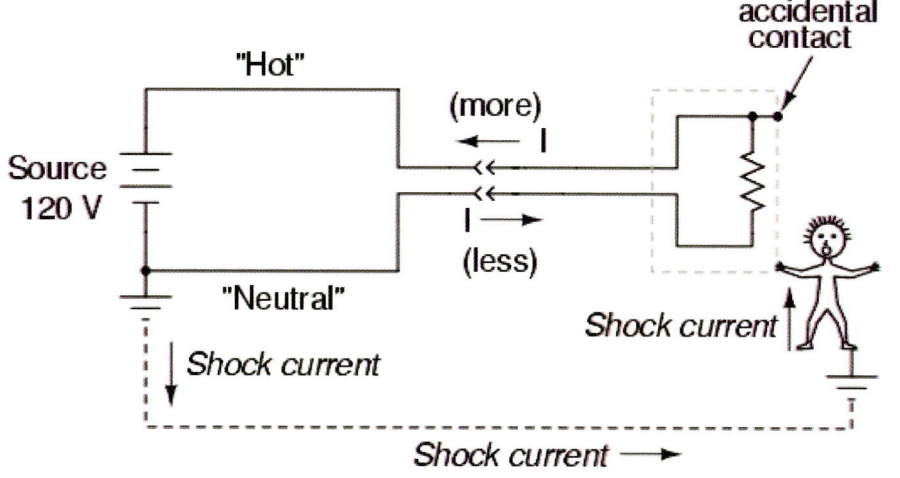

"...he was given a severe shock..."

AC Wiring (Warning)

But grabbing a hot lead with one hand and a neutral with the other, or the lead with one hand while standing in water, can send the current through the chest. One effect that electrical current has on the body is to make muscles contract, so a person getting a shock may be unable to release the item that's carrying the current.

The body isn't a perfect conductor of electricity but passing through the chest it takes only 0.05 amp to kill. That's barely enough to light a small bulb, and an amount which easily can pass through a human body that becomes a conduit between a hot AC wire and ground.
Two-prong plugs get put into sockets backwards (a condition known as reverse polarity). Circuitry chafes or cracks, exposing bare wire.

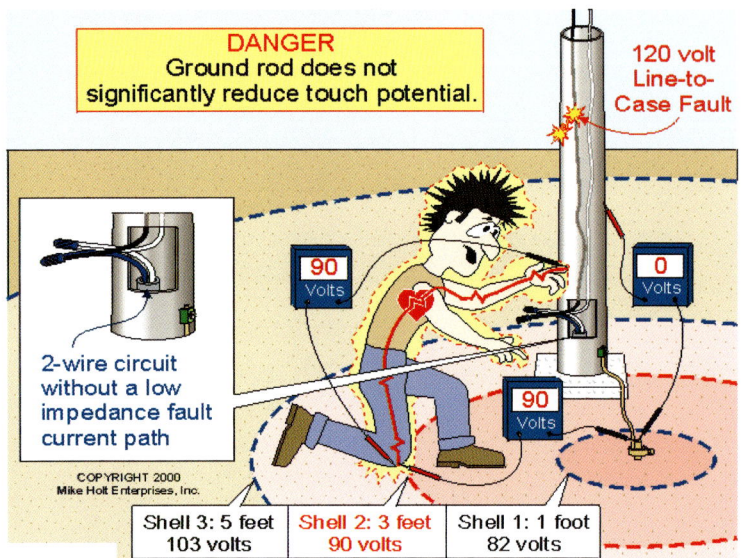

Running fuses continuously at full ratings

When matching circuit protection to the wire it protects, two facts contribute to the complexity of this task:
The amperage at which fuses actually blow, and circuit breakers actually trip, is considerably higher than their nominal ratings, the rating usually marked on the unit.
Wire and circuit protection devices heat up dramatically when they carry 100% of their rated value for several minutes or more.

AGC fuses, and most circuit breakers, blow or trip at about 130% of their rating. ANL fuses blow from 140% to as high as 266% of their rating. When fuses carry 100% of their rated current value, they generate excessive heat.
In combination, the heat produced by fuses and wires carrying high current can melt wire insulation and fuse blocks. This heat generation may become critical when loads run for a considerable time.

E.L.C.I

Equipment Leakage Circuit Interrupter is a residual current device which detects equipment ground fault leakage current and disconnects all ungrounded (110V & 240V) and grounded (110V Neutral)

Where is Located the ELCI ?

An ELCI shall be installed with or in addition to the main shore power disconnect circuit breakers.

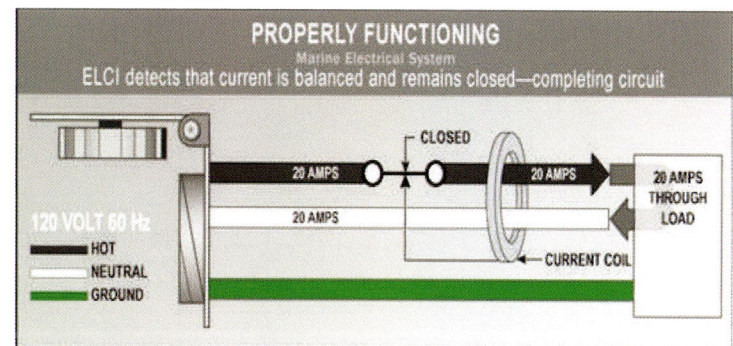

How ELCI Works?

In a properly functioning marine electrical system, the same amount of AC current flows in the hot and neutral wires
However, if electricity "leaks" from this intended path in these two wires to ground, this condition is called a ground fault. A good example of this is an insulation failure in the wiring of an appliance

Faulty Ground

A faulty ground can occur when the grounding path is broken through a loose connection or broken wire. For instance, a shore power cord ground wire may fail due to constant motion and stress.
In the bottom diagram, if the person just touch the bottom side of the resistor, nothing would happen even though their feet would still be contacting the ground.

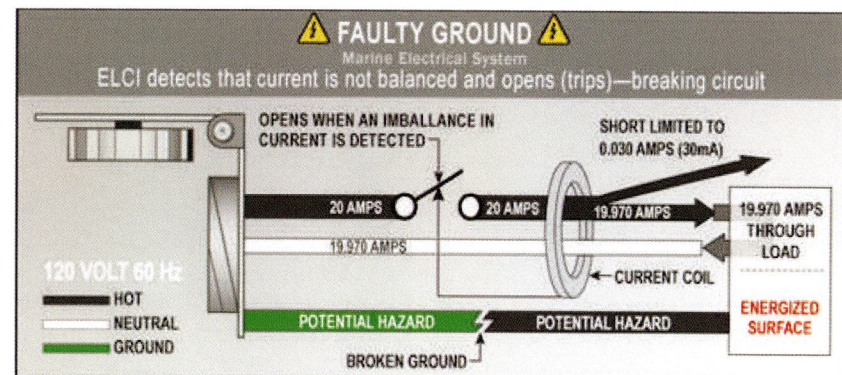

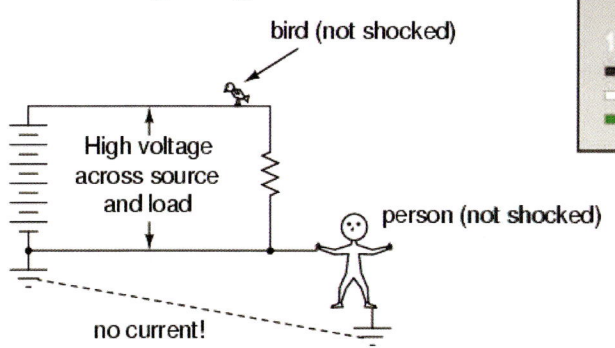

Shore Power Accessories
ELECTRICAL EQUIPMENT

Shore Power Cord

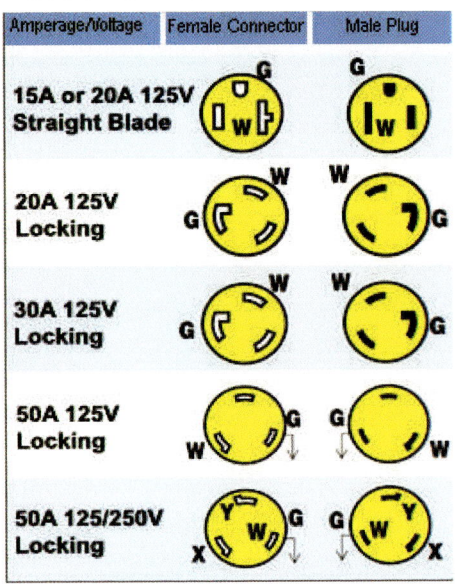

AC Power Pedestal

Episode 2: Dock Pedestals, wiring configuration & Troubleshooting

Scan this code to see the highlight video

In this video we will learn the procedure to repair and replace dock pedestals. In the video I explain the wiring procedure from the transformer to the pedestals and also the wiring internally to connect : Breakers Outlets, Power meters and measurement instruments.

Follow me

Shore Power Pedestal

The quality of marina wiring can vary greatly from place to place and from plug to plug.

Pedestal Connections

When polarity is incorrect or reversed, it is because the hot and neutral wires are switched. Without an indicator light you will not know when this condition exists because your electrical system still works normally.

The problem is that safeguards built into your boat's AC system will not function properly if the current is not flowing in the right direction.

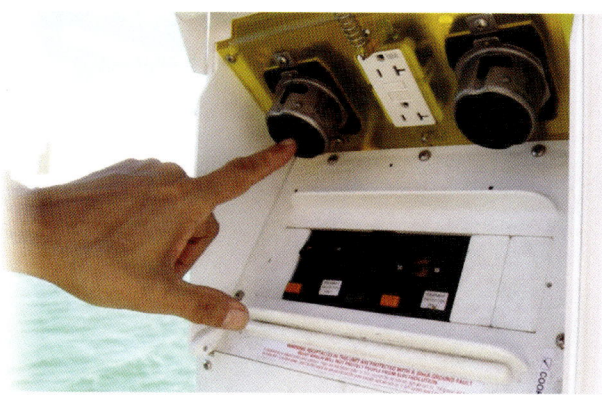

Your expectation is that the circuit is now dead and safe to touch. If the shore power polarity is reversed, you will likely be shocked and rudely surprised to find that the wiring is still hot. This can result in serious injury or death.

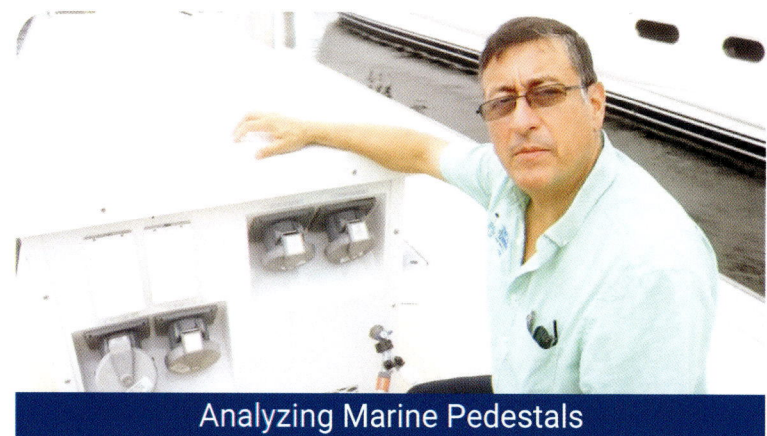

Analyzing Marine Pedestals

Chapter 12: AC Panel Board

Shore Power Cords

Sooner or later someone will use the shore power cable from the boat to the dock as a dockline.

Shore Power Cable Issues Diagnosis

Marina Dock and Pedestals Wiring Configuration

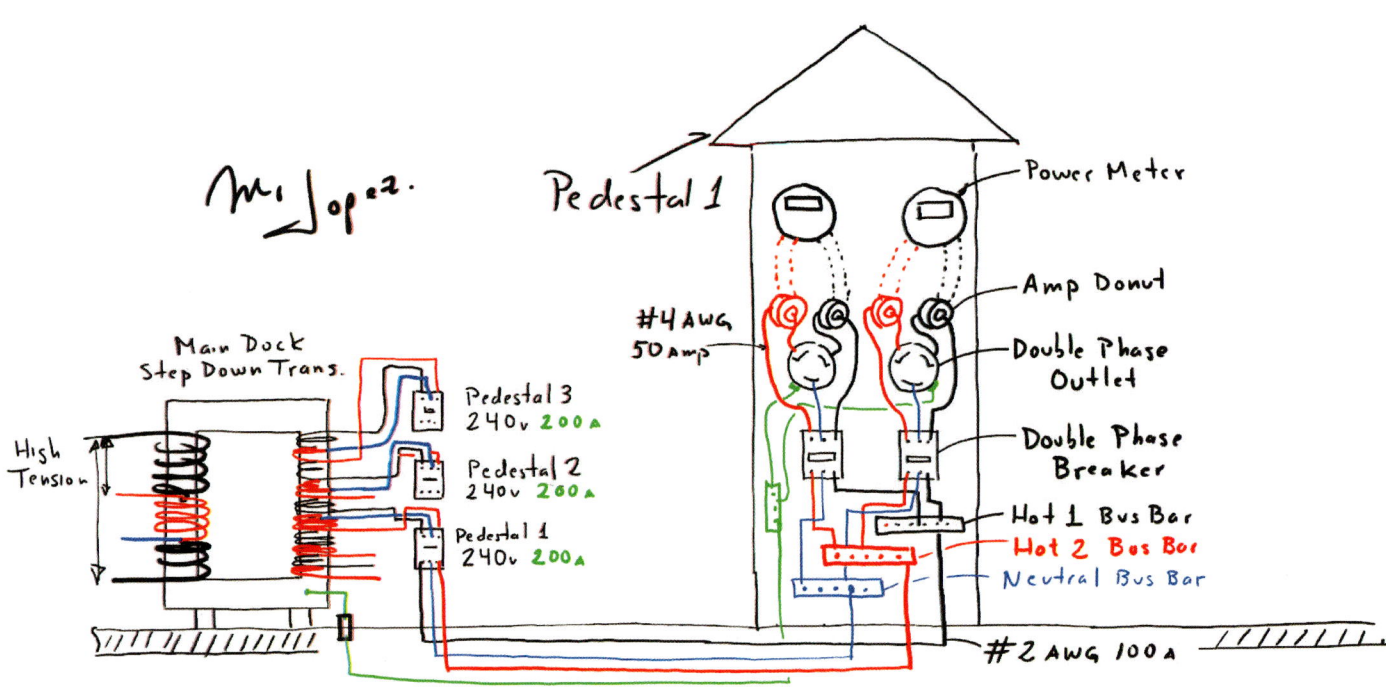

Shore Power Cable

The cable must be designated as flexible cord of type SO, ST or STO as show in the table.

Type	Description	Temp Rating
SO	Hard Service cord, oil resistant compound	140 F
ST	Hard service cord thermoplastic	140 F
STO	Hard service cord, oil resistant thermoplastic	160 F

Current-carrying conductors (Hot and Neutral) must be sized for the capacity of the shore Power circuit.

Conductor Size AWG	2 Conductors Max Amperage	3 Conductors Max Amperage
14	18	15
12	25	20
10	30	25
8	40	35
6	55	45
4	70	60

Shore Power Connectors

The shore end of the cable must have a locking and grounding cap with the proper male (plug) connector that matches the female shore receptacle.

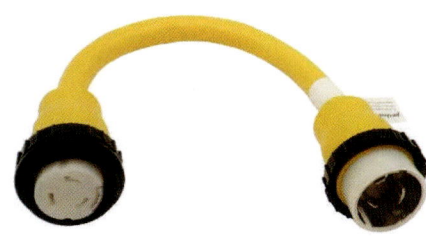

3 Phase Shore Power Pedestal

120/240V 3-Pole 4Wire Grounding

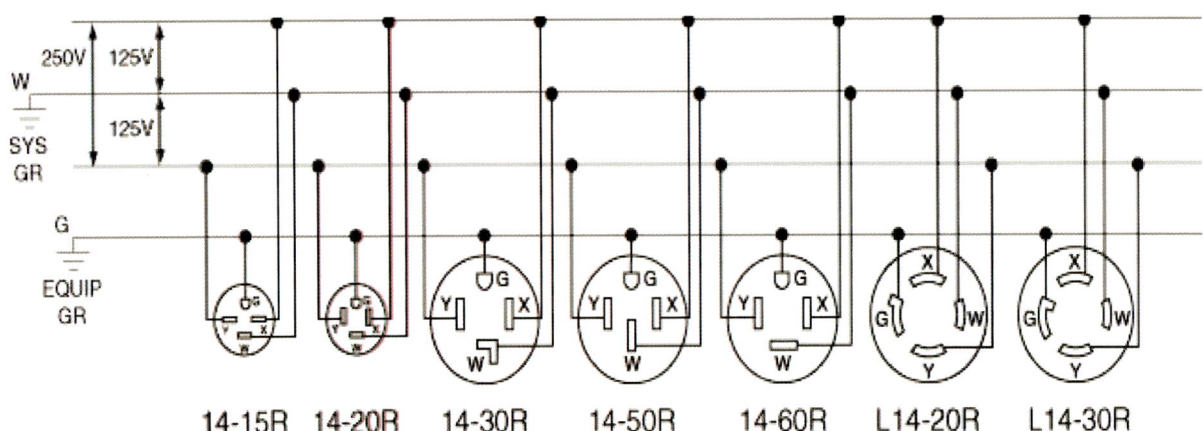

Shore Power Connection

A shore power connection is your ticket to a comfortable and smooth on board experience in most harbours.
Fixed alternating current (AC) is the power source to run appliances and to charge the batteries. This kit will make it easy for you to set up your own shore connection system.

120 Volt 30 Amp 120 Volt 50 Amp 120/240 Volt 50 Amp 120 Volt 15 Amp

What you should know about Shore Power Connectors

On connectors with white plastic centers, a little bit of discoloration around a contact is a probable indication of much more serious problems inside the connector. It's better to err on the cautious side, and if anything appears to be failing, replace the connector and the cord at the same time.

Recommendations

Electricity and water should not mix, and if they do, water will always win. With battery powered systems, if water gets to any connections or fittings, then the worst that can happen is that you have an inoperative system and a flat battery.

With shore supply voltages the risks from water are much greater. The voltages used can be lethal, so a sound installation and care in its use are vital.

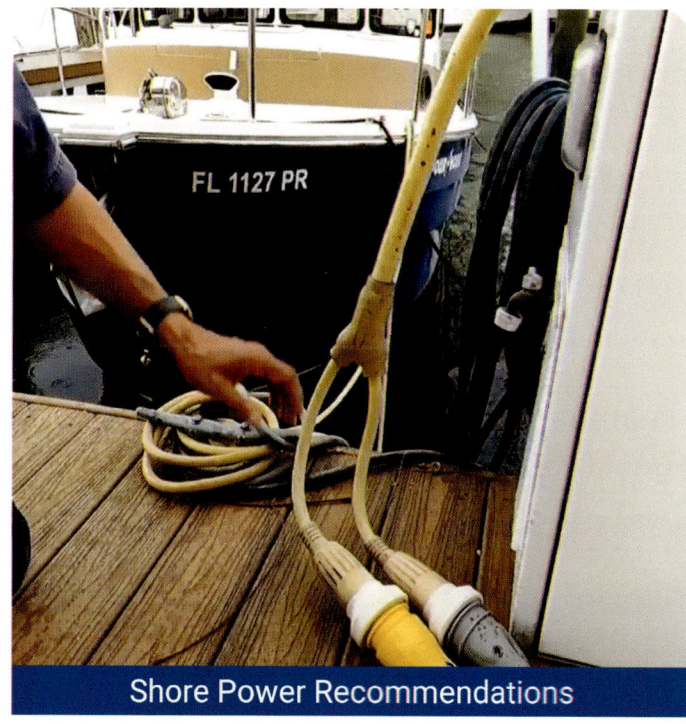

Shore Power Recommendations

What you should know about Shore Power Connectors

Cord should not be run directly from the boat to the dock without being secured to the vessel. If they are, the plug will be tugged every time the vessel moves, resulting in damage to the connector.

At the dock end, attach a string or other support to the cord near the connector to provide strain relief.

Ideally, connectors should not be located such that cords can be kicked or stepped on.

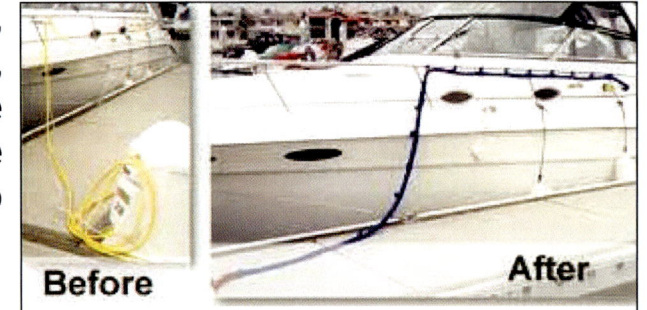

If this is not the case on your boat, and you cannot easily relocate the connectors, provide some means of tying the cord down to the deck or to an adjacent rail or fitting so it won't be damaged by an inadvertent kick – as well as to protect the inadvertent kicker!.

Side to side movement of the cord plug will stress the connector, and moisture entering the connection can allow corrosion to occur, increasing resistance, and causing heat/fire.

To prevent this, every cord should have a securing nut, and it should always be tightened when hooking up.

Improperly tightened screws attaching the wiring to the connector can cause an arc and generate heat.

Failure to turn off any loads within the boat (such as a running electric heater, water heater or refrigerator) prior to plugging in or unplugging the cord can cause an arc, pitting the contacts.

If this happens enough times a carbon film will form, creating a bad connection that will generate more heat.

Like many other components, shore power cables have a limited life and should be replaced well before they give out entirely. Just plain age and corrosion can lead to resistance/arcing/heat.

CHAPTER 13

AC Motors & Transformers

TOPICS

13.0 Single Phase Induction Motors	170
13.1 Dual Voltage Motors	171
13.2 Starting Capacitors	172
13.3 Three Phase Motors	175
13.4 Low and High Speed Motors	178
13.5 Three Phase Reverse Rotation	179
13.6 Contactors	179
13.7 Step Up and Step down Transformers	180
13.8 Y and delta Connection	182

Video Episode 1: How AC and DC Motors works & Reversible DC Motors

Scan this code to see the highlight video

In this video you will learn in a simple way how a DC or AC motor works. In the second part you will find information related with the wiring of a DC reversible motor using a reversible solenoid control box and a remote activation switch.

Follow me

Single Phase Induction Motor

When the power is applied to the stator, the coils on the stator produce electromagnetic fields.

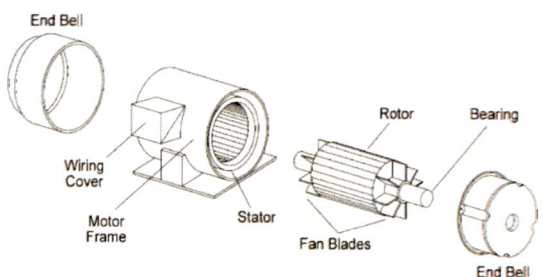

Single Phase Motors

To reverse the rotation, you must reverse the connections to the start windings. Not all the single-phase motors are designed for reversing rotation.

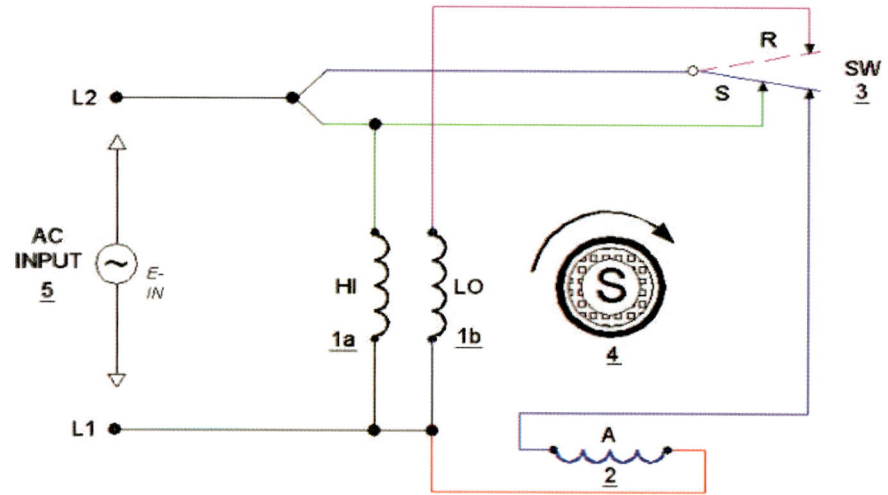

Induction Motors

The rotor revolves inside the stator housing. The stator induces an emf on the rotor, causing a current to flow through the bars, The bars connect at the ends of the rotor, completing the circuit and producing a magnetic field, that is converted in a torque.

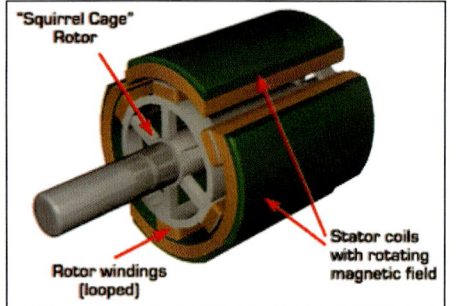

Double Winding

Start and Run winding at the Stator.

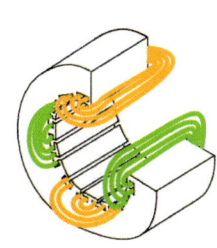 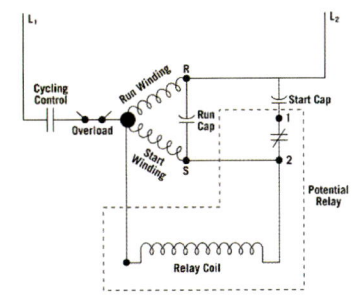 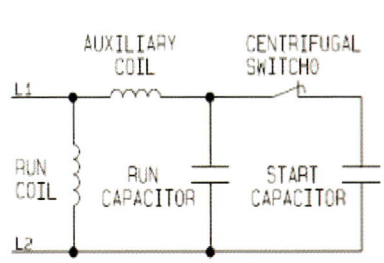

Dual Voltage Motor

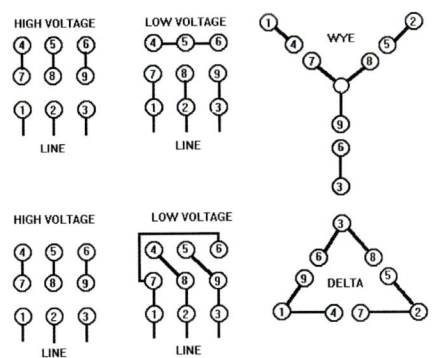

A dual Voltage Motor has the stator coils arranged in pairs of coils that allow the motor to be used with two different voltages.

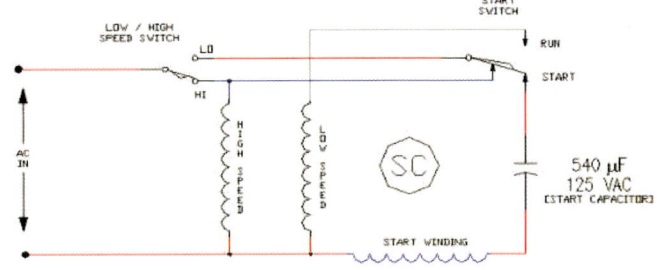

Note: Motor starts on "High Speed" winding regardless of speed switch setting

Starting Single Phase Motors

A single-phase current could be divided into a polyphase current and the starting problem would be solved.
This action is called **splitting phases**

TYPICAL SPLIT-PHASE AC INDUCTION MOTOR

Chapter 13: AC Motors & Transformers

Video Episode 2: AC Induction Motors & Reversing AC Motors with Contactors

In this video you will learn in a simple way how a DC or AC motor works. In the second part you will find information related with the wiring of a DC reversible motor using a reversible solenoid control box and a remote activation switch.

Follow me

Scan this code to see the highlight video

Start Capacitor

A **start capacitor** or **run capacitor**, is an electrical capacitor that alters the current to one or more windings of a single phase AC induction motor to create a rotating magnetic field.

Starting Capacitors

When an electrical motor is having trouble starting, such as an air conditioning compressor motor, blower motor, a refrigerator motor or a freezer motor, or even a fan motor, the repair technician may install a simple and inexpensive starting capacitor.

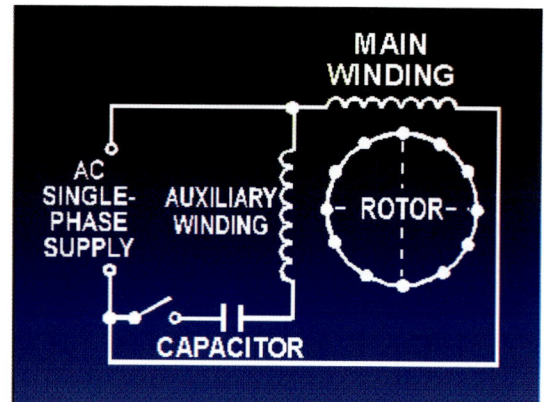
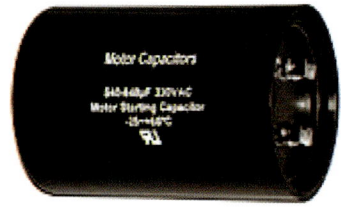

Chapter 13: AC Motors & Transformers

Starting Capacitors

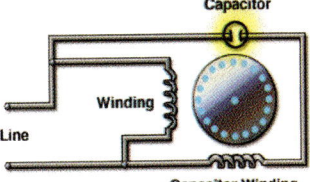

The starting capacitor is a simple electrical device which can give an extra voltage jolt or "boost" to get the hard-starting motor spinning.
Its function is to assist in developing high torque to get a motor turning over. It increases starting torque by aprox. 250%.
When replacing capacitors, one of equal value or more expressed in farads should be used.

Never use replacement with a lesser value, doing so may overheat the capacitor and / damage the motor.
In general starting capacitor is oval in cross section.

Splitting Phase With a Capacitor

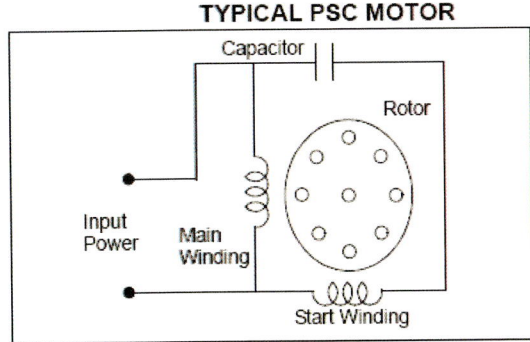

If a capacitor is connected in series with the starting winding, a much larger phase displacement can be created.
A larger phase displacement means that the motor will have a much higher starting torque than an induction start motor.

Start, Run Capacitor Motor

The two capacitors are connected to the motor windings, one capacitor to the start winding and the other to the run winding.

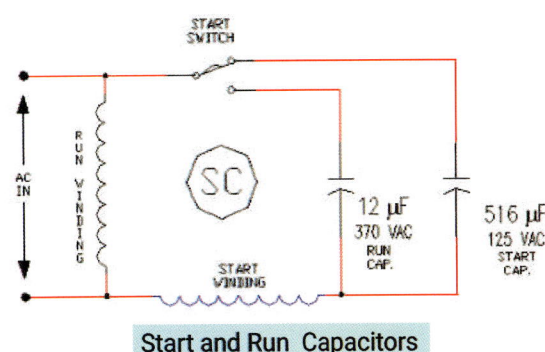

Start and Run Capacitors

Start Capacitors

Start capacitors above 20 microfarads (µF) are always non-polarized Aluminum Electrolyic Capacitors with non solid electrolyte and therefore they are only applicable for the short motor starting time.

Run Capacitors

Run capacitors are mostly polypropylene film capacitors and are designed for continuous duty, and they are energized the entire time the motor is running. [1] Run capacitors are rated in a range of 1.5–100 microfarads (µF or mfd), with voltage classifications of 370 V or 440 V.

Dual Run Capacitors

A dual run capacitor supports 2 electric motors, such as in large air conditioner or heat pump units, with both a fan motor and a compressor motor.

Round dual run capacitors (shaped as round cylinders) are commonly used for air conditioning, to help in the starting of the compressor and the condenser fan motor.

Start, Run Capacitor Motor

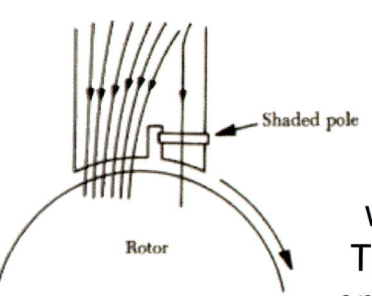

The rotor is squirrel cage type. A slot is cut in the face of each pole, and a single turn of heavy wire is wound in the slot.

As the current rises in the first quarter cycle of the AC wave, a magnetic field is formed in the field winding.

This interaction induces a current that opposes change in the magnetic field of the field coil.

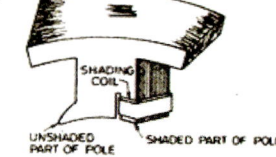 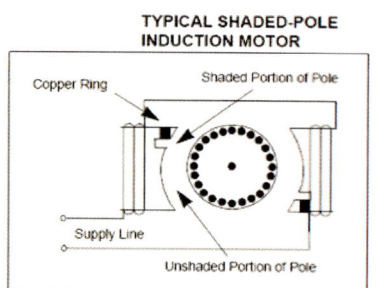

Three Phase Motor

When the three phase power is applied to the stator, the coils produce electromagnetic fields, that create rotation.

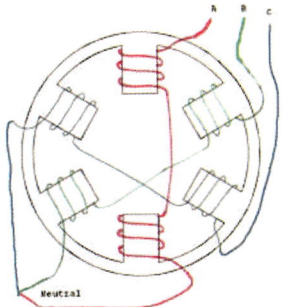

Three-phase electric motors are more efficient than single-phase motors and consume less current for a given horsepower rating. Because a three-phase motor requires less amperage than a single-phase motor of the same horsepower rating, the motor can be wired using smaller size conductors.

Three Phase Motors

Most industrial motors are three phase. The main reason for this is that there is very little maintenance of a three phase. Industrial motors do not have the starting devices that single-phase motors have.

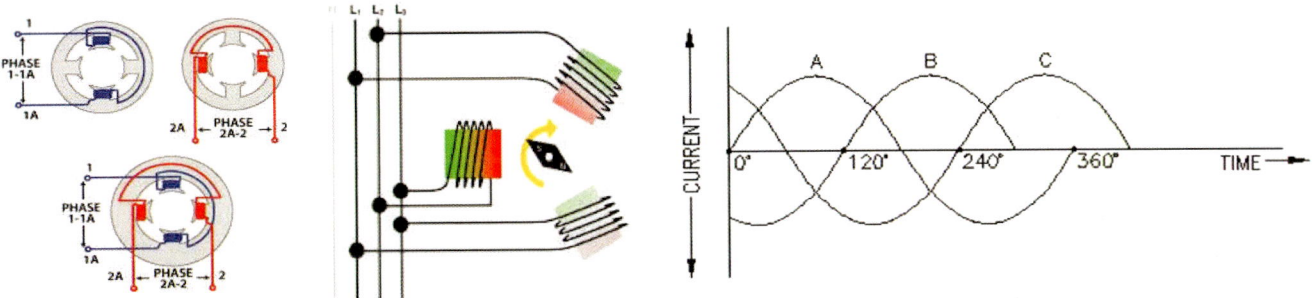

The three phases of alternating current that supply power for the motor produce the phase shift needed to get the motor started and to keep it running once it is started. All commercial power generated in the United States is generated as three phase.

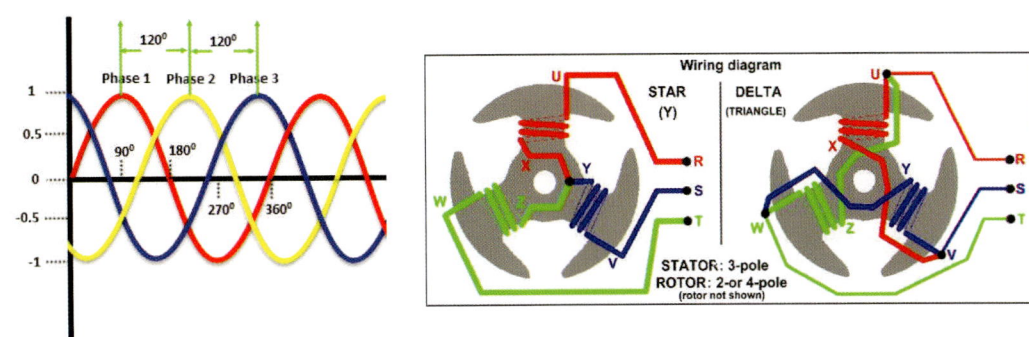

Three Phase Wiring

The National Electric Code does not specify specific conductor colors for three-phase current, it is common to use **black**, **red** and **blue** wires to identify lines **L1**, **L2** and **L3** respectively.

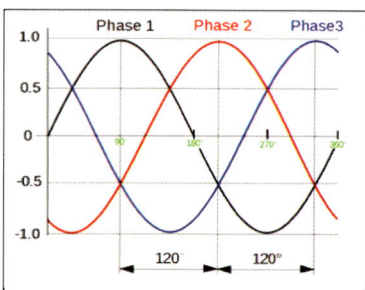

The voltage cycle of each line lags its predecessor by 120 degrees -- L2 reaches its peak voltage after L1, and L3 reaches its peak voltage after L2.

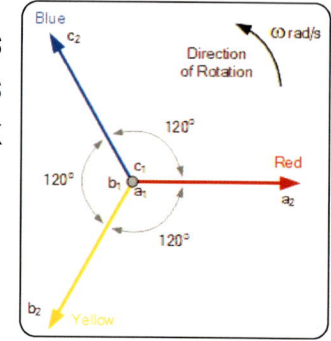

Two wiring configurations, Wye and Delta, indicate the wiring methods for three-phase motors.

Wye Connection

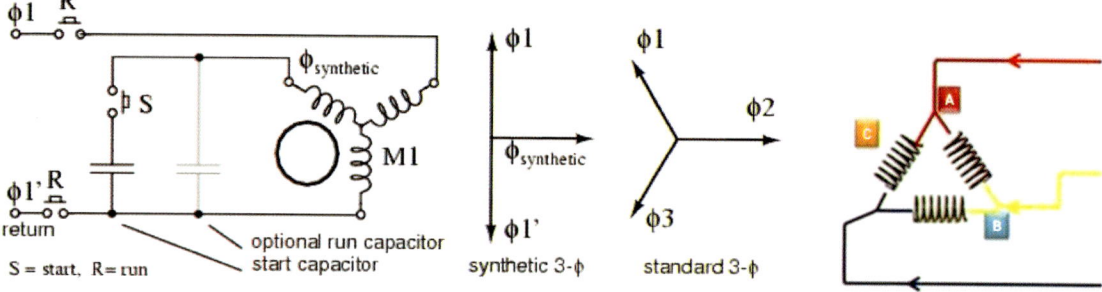

Single Vs. Three

SINGLE PHASE MOTORS				
Motor H.P	Input Voltage	Full Load Amperes	Breaker Size	Min. Copper Wire Size
3/4	230-240	6.9	15	14
1	230-240	8	15	14
3	230-240	17	35	12
5	230-240	28	60	10
7 1/2	230-240	40	80	8
THREE PHASE MOTORS				
Motor H.P	Input Voltage	Full Load Amperes	Breaker Size	Min. Copper Wire Size
3/4	230-240	3.2	15	14
1	230-240	4.2	15	14
3	230-240	9.6	20	14
5	230-240	15 1/15	30	14
7 1/2	230-240	22	45	10

Three Phase Motors

Three-phase power requires at least three, and sometimes four, wires for proper distribution.

Three phase motors (3ø) are perfect for machine-tool and general uses where dust and dirt are prevalent.

Three-phase motors are available in sizes of ¼, 1/3, ½, and ¾ horsepower. They may be used for pumps, compressors, fans, blowers, conveyors, farm machinery, saws and machine tools.

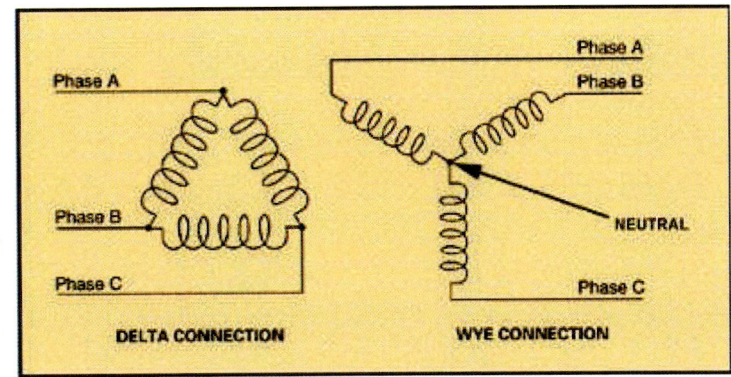

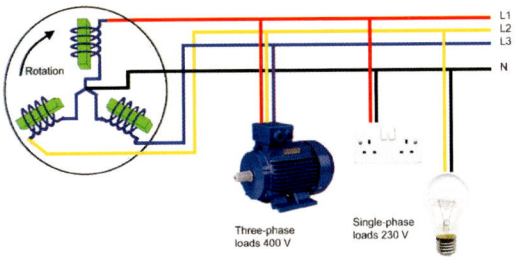

They also produce less vibration and noise than single-phase motors, making them suitable for applications where noise and vibration can be a problem.

How are produced the three phases

Basically, the power company generators produce electricity by rotating (3) coils or windings through a magnetic field within the generator.

These coils or windings are spaced 120 degrees apart. As they rotate through the magnetic field they generate power which is then sent out on three (3) lines as in three-phase power.

Three Phase Motors Analysis

Three Phase Power Efficiency

By the use of three conductors a 3 phase system can provide 173% more power than the two conductors of a single-phase system.
When 3-phase AC power is applied to the stationary electromagnet coils, then the rotor will rotate.
Three-phase power allows heavy duty industrial equipment to operate more smoothly and efficiently. 3 phase power can be transmitted over long distances with smaller conductor size.

Motors, Generators & Transformers

Motors, generators, and transformers are similar in that their basic principle of operation involves induction.
The premise for motor operation is that if you can create a rotating magnetic field in the stator of the motor, it will induce a voltage in the armature that will have magnetic properties causing it to 'chase' the field in the stator.

Question

What would we have to do in order to reverse the rotation of this three-phase induction motor?.
Answer: Reverse any two lines. This will reverse the phase sequence (from RST CW to SRT CCW).

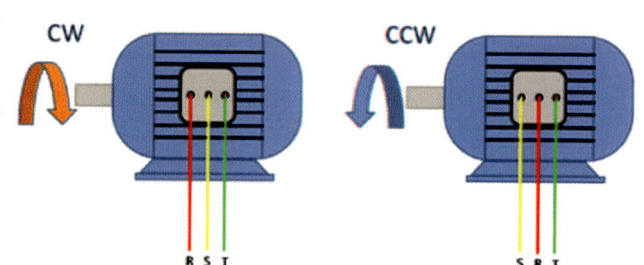

Low & High Speed

Some AC induction motors are equipped with multiple windings so they may operate at two distinct speeds (low speed usually being one-half of high speed).
The difference between the two connection schemes is the polarity of three of the coils in relation to the other three. This is called the consequent pole design of two-speed motor, where you essentially double the number of poles in the motor by reconnection.

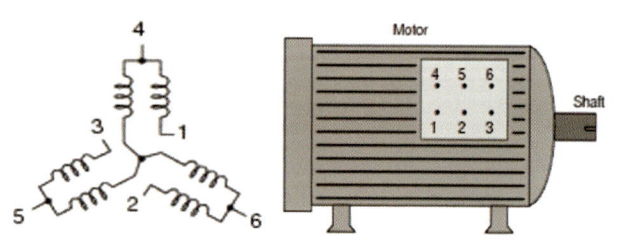

Speed	φ-A	φ-B	φ-C	Left open	Shorted together
Low	1	2	3	4,5,6	
High	4	5	6		1,2,3

Three Phase Reverse Rotation

The properly way to reverse rotation in a Three phase motor is by means of "Contactors"

Contactors

A contactor is a remotely activated breaker.
The remote activation could be a low current AC or DC signal.

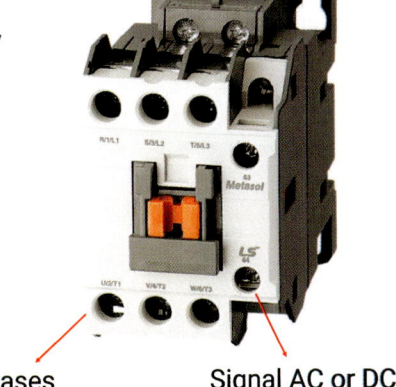

Phases — Signal AC or DC

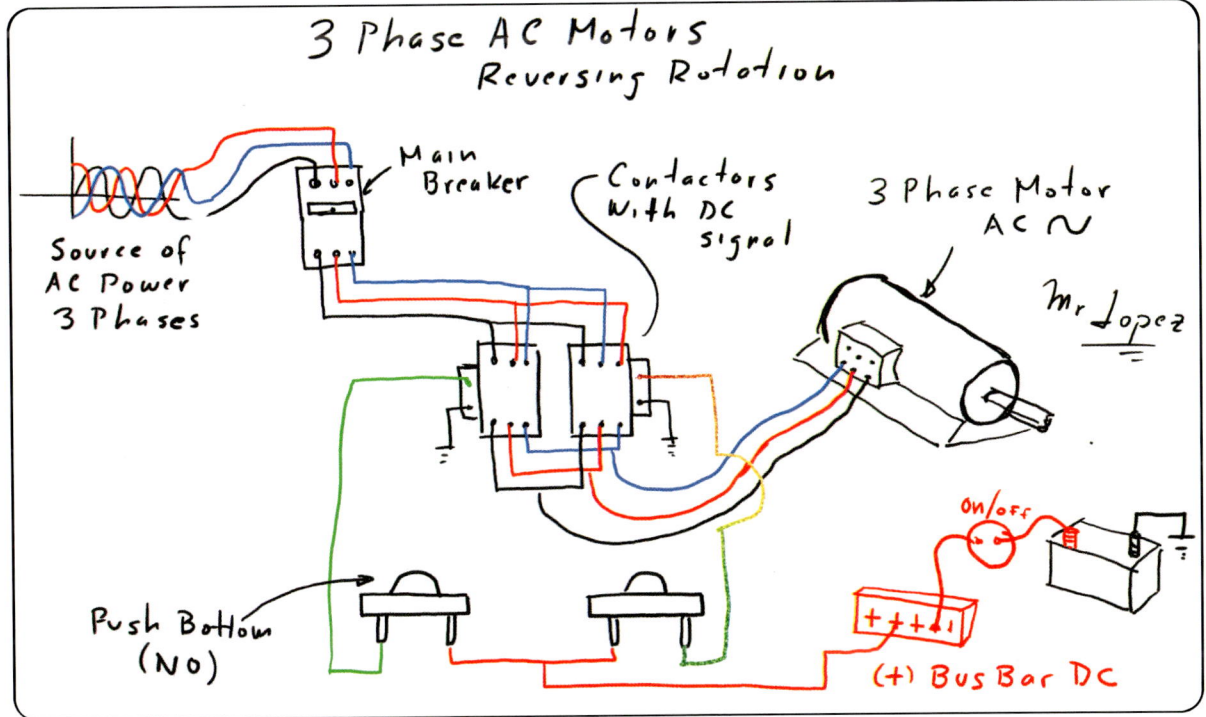

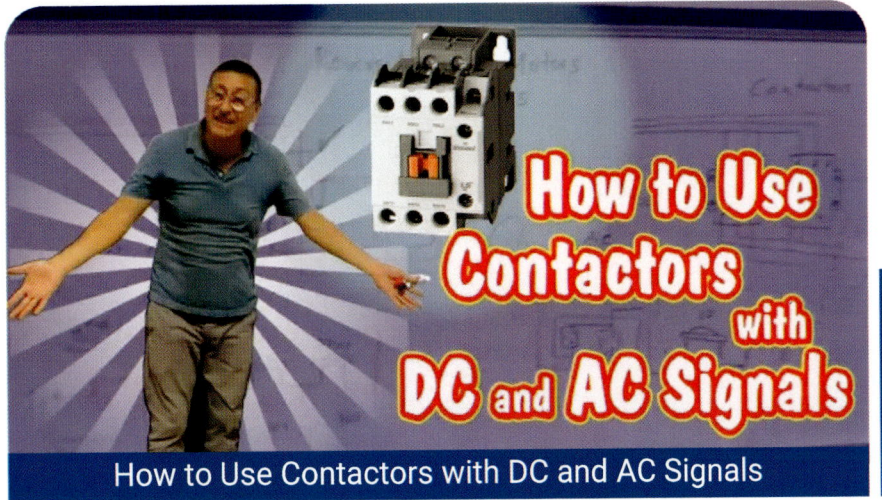

How to Use Contactors with DC and AC Signals

Transformers

Video Episode 3: Transformers Step-Up, Step-Down, Isolator Transformers

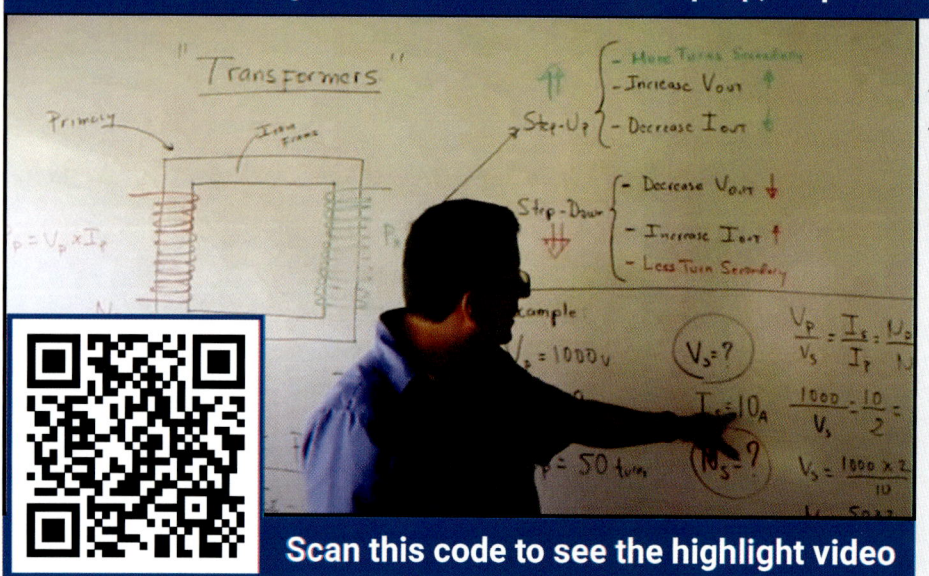

Scan this code to see the highlight video

In this video you will learn about transformers and how to calculate the capacity of transformers single phase, double phase and three phases.

Follow me

Step Up Transformers

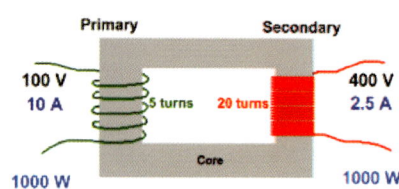

Depending of the number of turns in the secondary coil, the output voltage can be increased or decreased.

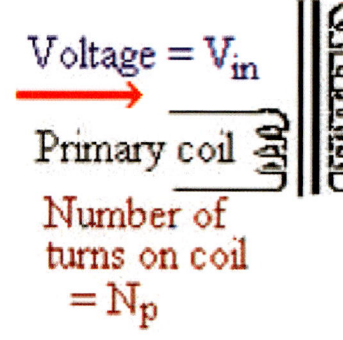

If N_s is more than N_p, we have a "Step-Up" transformer

If N_s is less than N_p, we have a "Step-Down" transformer

All of these questions can be done using $\dfrac{V_{in}}{V_{out}} = \dfrac{N_p}{N_s}$

Step Down Transformers

If the purpose is to increase voltage, the secondary coil must have more turns and therefore a thinner wire. Note that at best the output power equals the input power.

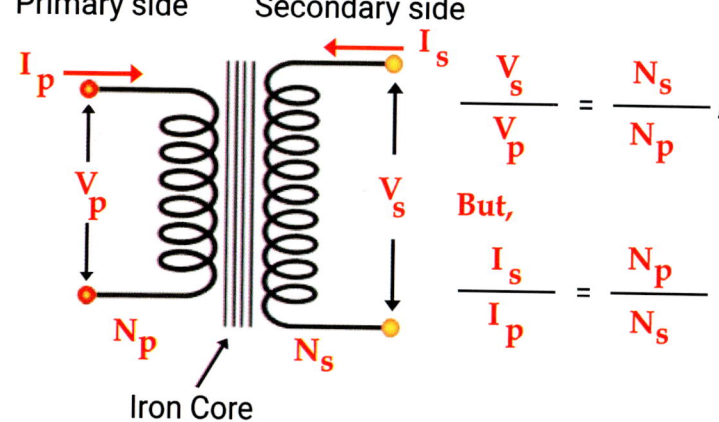

$$\frac{V_s}{V_p} = \frac{N_s}{N_p},$$

But,

$$\frac{I_s}{I_p} = \frac{N_p}{N_s}$$

Ideally, $P_{out} = P_{in}$ .or $V_s I_s = V_p I_p$

Writing as proportions: $\frac{V_s}{V_p} = \frac{I_s}{I_p}$

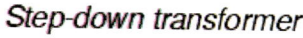

This is true for an ideal transformer only.

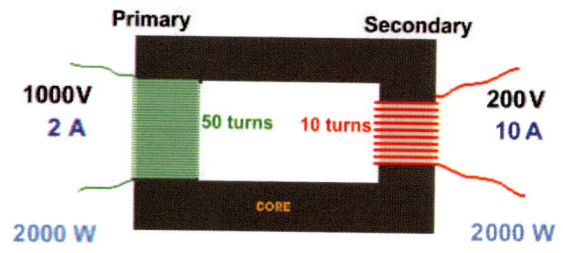

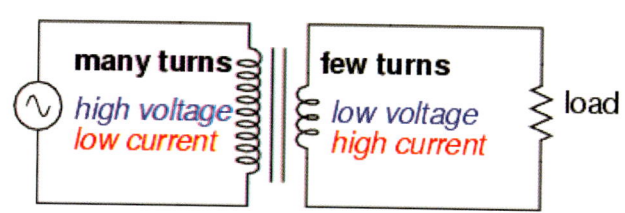

Isolator Transformer

If both coils have the same number of turns the transformer is called an isolator transformer (See Chapter 10 - Bonding System).

In Marine applications each day the use of the isolator transformer is critical due that the output signal is clean and safe.

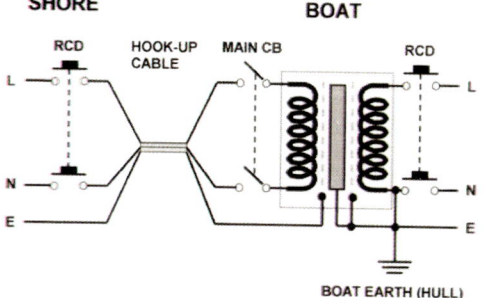

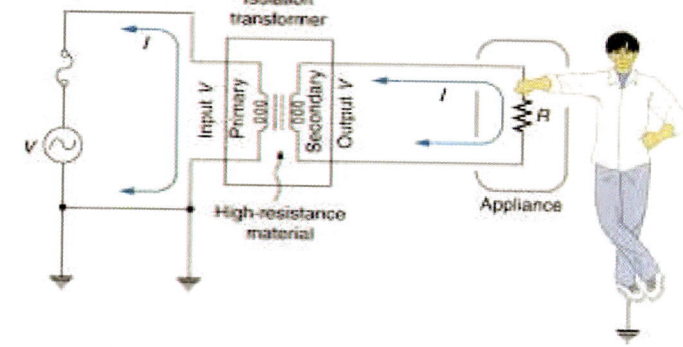

Three Phase Winding

Some transformers are designed for two and three phases.

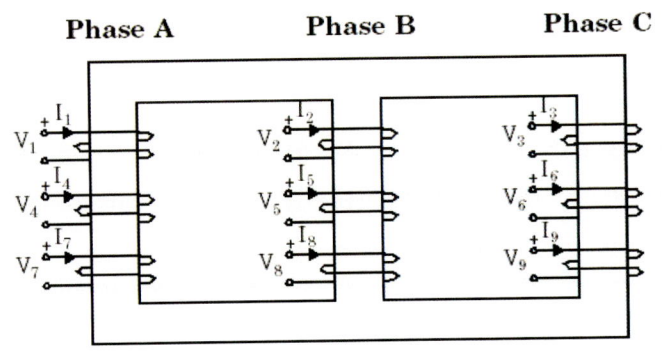

Three-phase winding 1
Three-phase winding 2
Three-phase winding 3

Delta – Y Transformer

Delta-wye (Δ-Y) transformer is a type of three phase electric power transformer design that employs delta-connected windings on its primary and wye/star connected windings on its secondary. A neutral wire can be provided on wye output side.

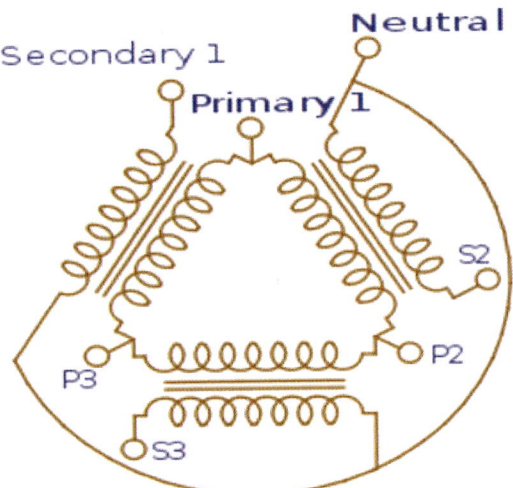

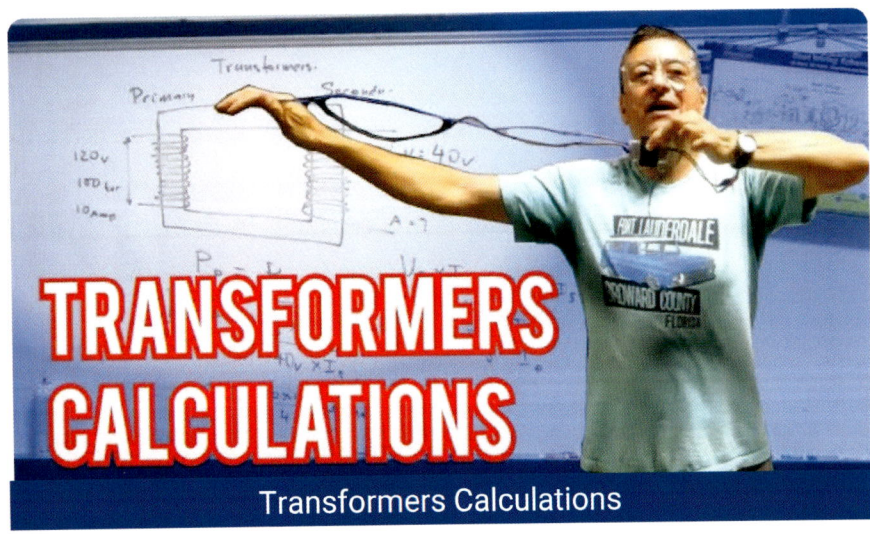

Transformers Calculations

CHAPTER 14

Marine Generators

TOPICS

14.0 AC Generator Theory	184
14.1 Types of Excitation	186
14.2 AC Alternator	186
14.3 The frequency	188
14.4 Voltage Regulator	190
14.5 The Megger	193
14.6 How to Install a Marine Generator	195
14.7 Troubleshooting	197
14.8 The Generator Won't Crank	197
14.9 The Generator start but Stop Suddenly	198
14.10 The Generator Crank but Doesn't Start	199
14.11 Low Frequency	200
14.12 Under Voltage	200
14.13 Design the Harness	201

Video Episode 1: Generators fundamentals & Conversion from 50 Hz to 60 Hz

Scan this code to see the highlight video

In this video you will learn about the fundamentals of marine generators, the quality of the signal and how to convert a 50 Hz genset into a 60 Hz.

Follow me

Chapter 14: Marine Generators

Words to Know

Alternating current (AC): Electric current in which the direction of flow changes back and forth rapidly and at a regular rate.
Armature: A part of a generator consisting of an iron core around which is wrapped a wire.
Commutator: A slip ring that serves to reverse the direction in which an electrical current flows in a generator.
Efm = Electro Magnetic Force = Voltage.

Theory

Electric generators are constructed from current-carrying loops that continuously rotate in a magnetic field.
To make the magnetic field stronger, the loops are wrapped around an iron core called an **armature.** (The Armature is part of the Rotor).
The loops are constantly rotating because the current inside of them changes direction, which causes an induced emf to be generated. An emf is not a force, but a voltage.

AC Generator Theory

Two types of AC generators are:

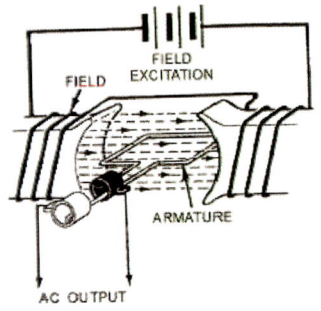

A ROTATING ARMATURE ALTERNATOR

1.- **"Armature Type"** or (Generator Type or Self excitation type). It is option A; in which the output AC is generated by the rotor and the excitation is produced on the stator with DC current coming from a bridge of diodes, who receive and small AC signal produced during the engine cranking due to the weak magnetic signal, stored in the rotor frame.

Marine Generator Fundamentals Part I

AC Generator Theory

2.- **"Alternator Type"**, It is option **(B)**, in which the output is generated by the stator whit excitation DC coming from the ignition switch through the brushes.
Both types of Generator are commonly used in marine applications. The strength of the AC output in both cases depends of: The voltage and intensity (DC) on the exciter, the thickness of the wire on the coils and the number of turns in each coil.

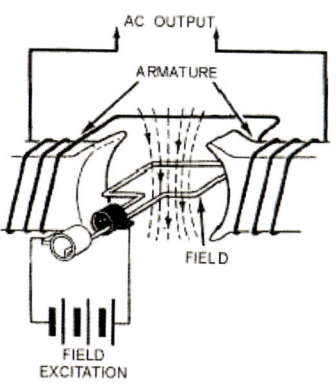

B ROTATING FIELD ALTERNATOR

AC Alternator

The excitation (Input) is by means of DC Current on the rotor through the brushes.

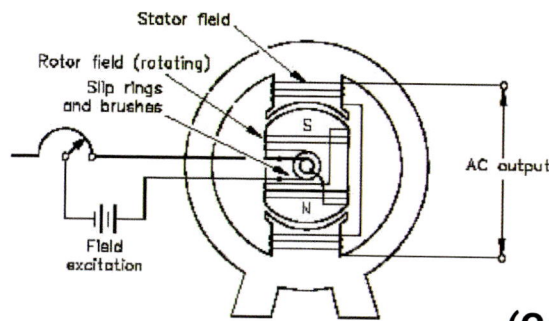

One end of the rotor is bolted at the engine flywheel and the other end receive the excitation (DC) from the ignition switch through the brushes. The output AC current is produced at the stator coil, then the AC signal pass through the voltage regulator to adjust: Amplitude (**Volt**), Stabilization (**Stab**) and Gain (**Gain**). In some new designs also the Frequency (Only fine adjustment) can be adjusted at the voltage regulator.
The output signal coming out from the voltage regulator pass through the AC switch selector and feed the AC panel.

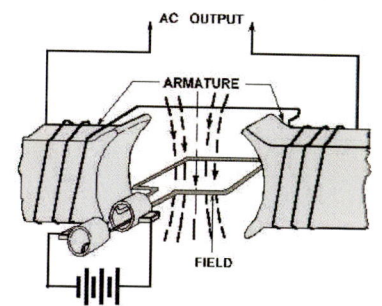

Marine Generator Fundamentals Part II

Types of Excitation

An exciter is part of the generator package supplying direct current to the alternator field windings to magnetize the rotating poles (In alternator Type) or to magnetize the stator poles (In the Armature type or self excitation type).

A poor AC output signal is normally caused for a weak DC excitation.

In a Generator, Alternator Type the brushes pass the DC current coming from the ignition switch.

In a Generator, Armature type there are two sets of brushes. One set to pass the AC current produced on the rotor to the voltage regulator then to the main AC switch selector. And the second one to pass the weak AC signal produced at the beginning during the cranking process to feed the bridge of diodes to produce the self excitation at the stator coil.

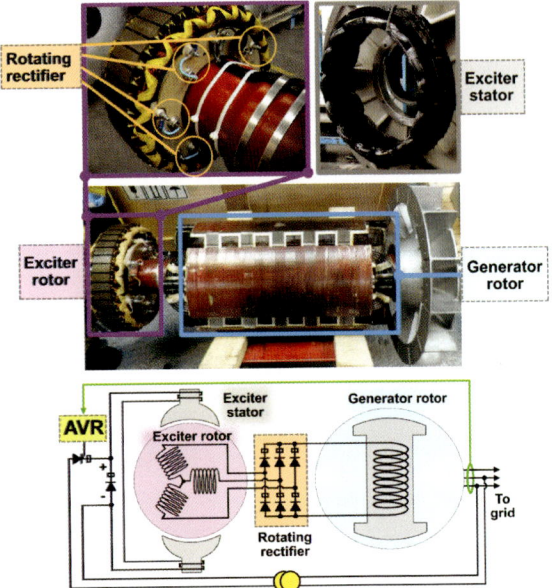

AC Alternator

The strong magnetic field is produced by a current flow through the field coil of the rotor.
The field coil in the rotor receives excitation through the use of slip rings & brushes.
Two brushes are spring-held in contact with the slip rings to provide the continuous connection between the field coil and the external excitation circuit.

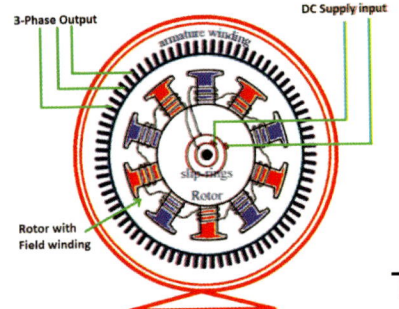

The armature is contained within the windings of the stator and is connected to the output.

Each time the rotor makes one complete revolution, one complete cycle of AC is developed.

A generator has many turns of wire wound into the slots of the rotor.

The magnitude of AC voltage generated by an AC generator is dependent on the field strength, the thickens of the wire coil, the number of turns and speed of the rotor.

DC Input excitation through the brushes

AC Generator (Armature Type)

The excitation occurs in the stator by means of a bridge of diodes that rectify an small AC current (Produced by the generator during the cranking), into DC current. This procedure is called "self-excitation", because the generator produce your own current of excitation.

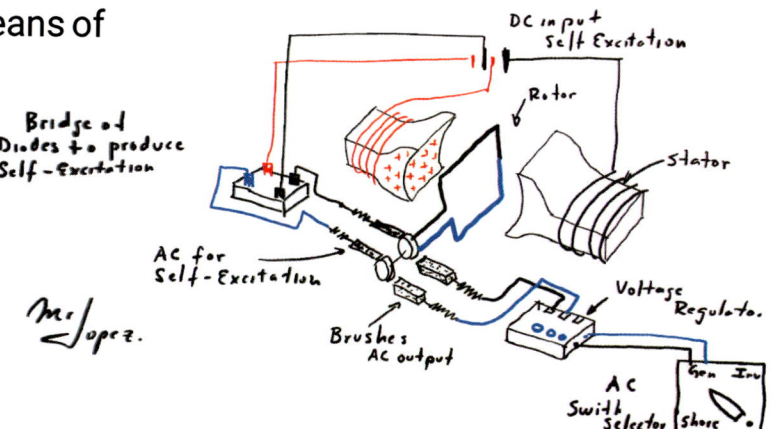

DC Current rectified by the bridge of diodes to feed the stationary magnets on the stator

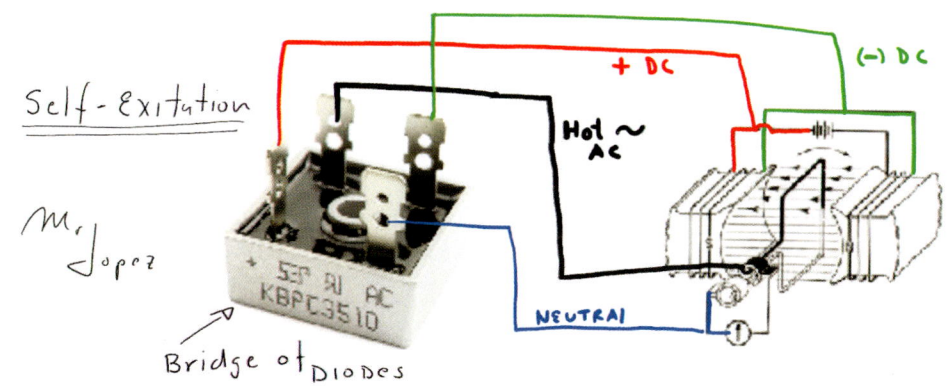

AC Current created by the generator during the cranking process

AC Alternator   Vs   AC Generator

- Excitation on the rotor by means of the brushes and the slip rings.
- The magnets are on the Rotor.

- Self-excitation on the stator by means of a bridge of diodes that receives the AC current produced by the rotor during the cranking process through the split rings and the brushes.
- The magnets are on the stator

The Frequency

The frequency of the electrical system varies by country; most electric power is generated at either 50 or 60 Hz. Some countries have a mixture of 50 Hz and 60 Hz supplies. Frequency is always measured and expressed in hertz.

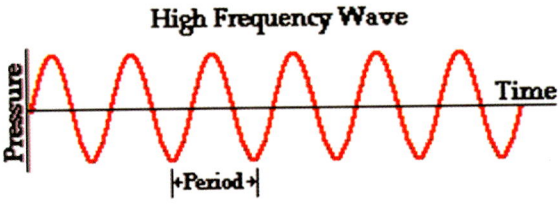

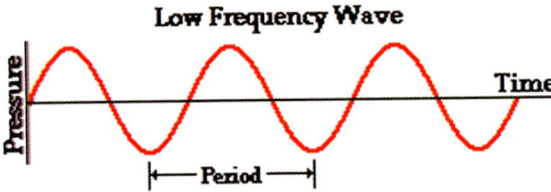

Frequency Vs. RPM

How to Check Frequency

The frequency of the generated voltage is dependent on the number of field poles and **the speed** at which the generator is operated. Frequency, measured in Hertz (Hz), is the number of complete cycles per second in alternating current direction.

When a Direct Current (DC) voltage is applied to the field windings of a dc generator, current flows through the windings and sets up a steady magnetic field. This is called Field Excitation.

Chapter 14: Marine Generators

AC Generator Frequency

Generator Frequency (f) = Number of revolutions per minute of the engine (N) * Number of magnetic poles (P) / 120.
For a 4-pole generator, an engine speed of 1,800 rpm produces output of 60 Hz. Reducing the engine speed to 1,500 rpm yields an output of 50 Hz.

Generator Frequency (f) = Number of revolutions per minute of the engine (N) * Number of magnetic poles (P) / 120
For a 4-pole generator, an engine speed of 1,800 rpm produces output of 60 Hz. Reducing the engine speed to 1,500 rpm yields an output of 50 Hz.
Generator controller units undertake real-time monitoring and control of your unit. Built-in protective functions automatically shut down your generator in the case of excess engine rpm or very low output frequency.

Frequency Calculator

The frequency of the generated voltage is dependent on the number of field poles and the speed at which the generator is operated.

$$n = f = \frac{NP}{120} = Hz$$

f = Frequency (Hz) ; N = rpm; P = # of Poles
120 = Conversion from min to sec and from poles to pole pairs.

$$\frac{60 \text{ seconds}}{1 \text{ minute}} \times \frac{2 \text{ poles}}{\text{pole pair}}$$

Chapter 14: Marine Generators

Voltage Regulator

Video Episode 2: Voltage Regulator Adjustment and Megger test procedure

In this video you will find valuable information related with the calibration of parameters such as: Voltage, Frequency, Stabilization, Gain in a typical marine generator using the Voltage Regulator Board.

Follow me

Scan this code to see the highlight video

How a Voltage Regulator Works

The "voltage regulator" work by stabilizing the output voltage of the back-end at variable loads, and helps the generator respond to overloads.

Voltage Regulator Configuration

In a typical voltage regulator board, there are three or four calibration Controls:

- **Volt.**
- **Stab.**
- **U / F**
- **Gain**

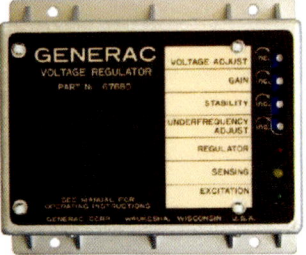

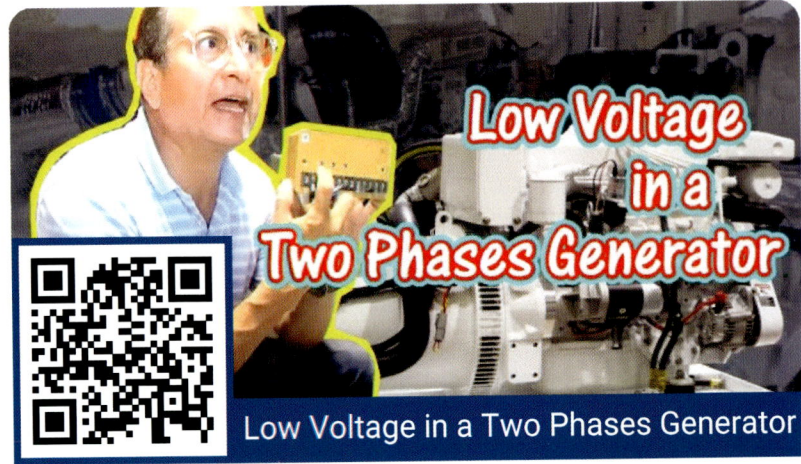

Low Voltage in a Two Phases Generator

Output Voltage Calibration Control "Volt".

This calibration button should be used to set adjustments in the range of 10% output voltage, per phase up & down.

If you find an output voltage of 80 V per phase in a 60 Hz generator. If that difference is more than 10%, then when you try to correct the deviation, the capacitors could burn out.

Coarse Output Voltage Adjustment

If the differences in output voltage are greater than 20%, the following verifications must be made:

- The neutral line must be checked as well as its connection to the ground point.
- The quality of the stator and rotor windings should be checked with the megger.
- The excitation quality must be checked (Brushes and /or Rectifier condition).

Voltage Regulator NO Adjust

"Stab". Calibration Control

This adjustment Control can be used when the output voltage shows instability and small fluctuations.

These adjustments are made with the generator running at full load applied and with an Analog Voltmeter connected to the main breaker output.

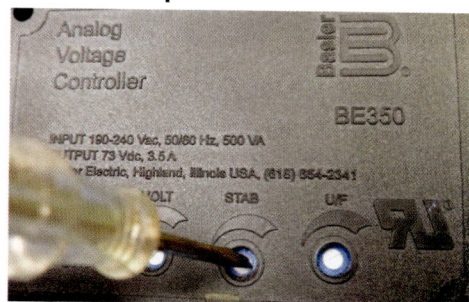

"U/F" Calibration Control

Again, the adjust button for frequency calibration should be used only for fine adjustments. Two or three Hz above or below.

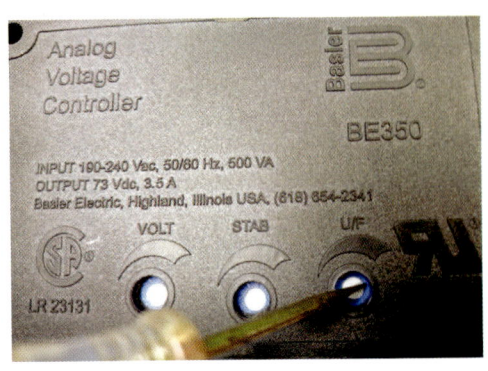

Coarse Frequency Adjustment

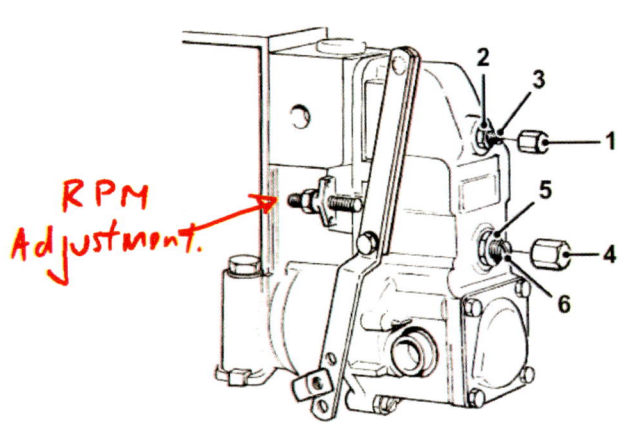

When the frequency setting is greater than 5 Hz this must be done directly on the governor at the fuel injection pump.

How to Avoid Frequency Modifications

To avoid alterations in the generator frequency, the idle bolt adjustment and WOT bolt adjustment should be locked with a safety wire using this technique.
(See video clip)

"Gain" Calibration Control

The gain control is a one-turn potentiometer. It is used to adjust the sensitivity of the governor.
The Gain control is used to adjust the response of the generator to recover RPM once a load is applied.

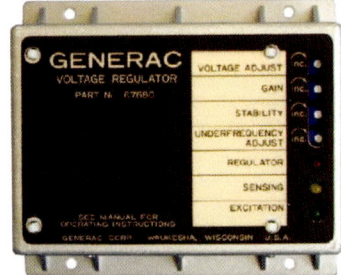

The Megger

Megger has become the generic description for a high voltage, low current insulation tester. The word is short for megohm-meter.
Although any Ohmmeter or Multimeter may appear capable of similar measurements, only a Megger type instrument can test the quality of the insulation at or above its operating voltage.

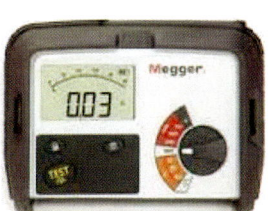

How Do Megohmmeters Work?

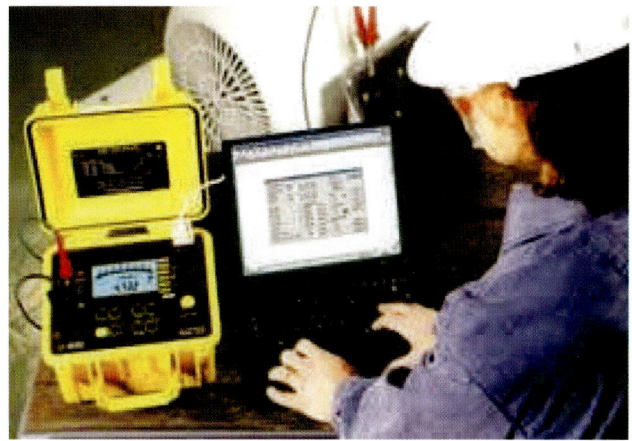

The megger consists of a DC generator and a direct reading ohm meter. The megger is a measuring device that tests high electrical resistance. It is used to check the winding condition in any type of motor, alternator, or generator.

Typically these measurements are made on electrical wires and motor windings to test the insulation value of the wires.

The readings are measured in millions of Ohms or Mega-Ohms.

Instructions

Connect one of the megger's leads to the electrical frame or earth ground of the electrical system. In testing motor windings, this lead will be connected to the actual metal frame of the motor.

Attach the other lead from the meter to the bare copper end of the wire or one of the motor terminals. Check to be sure the other end of the wire being tested is in free air or covered by an insulator such as a piece of tape or wire nut.

Chapter 14: Marine Generators

Instructions

Turn the meter on or begin to crank the generator handle. It may take 2 to 5 seconds for the high voltage to build inside the wire or motor windings.

 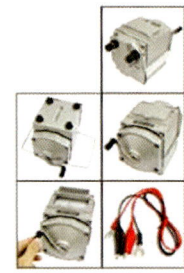

A reading of greater than 999 megohms is a near-perfect resistance reading for a wire or new motor. Resistance readings less than 1.5 megohms may present problems in old wires or used motors. In most cases, any reading in between these values may be fine, unless trouble has been experienced in these circuits.

 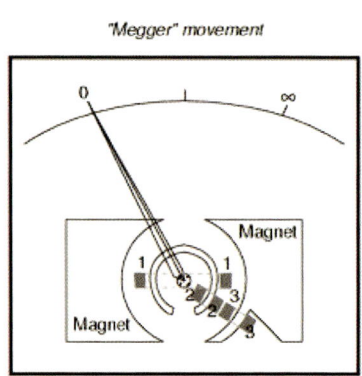 

How to use the Megohmmeter

Chapter 14: Marine Generators

Video Episode 3: How to install a Marine Generator

Scan this code to see the highlight video

In this video you will learn how to install a marine generator following ABYC recommendations and industrial practices. The video follow a detailed process with some tips and recommendations.

Follow me

How to install a Marine Generator

To Install a new generator successfully follow these steps:
1.- Size the capacity of the generator in Kw according to the capacity of the AC Panel in Kw, plus 20% for the safety factor.
2.- Analyze different places in the engine room to install de generator, avoiding :
- - Proximity to propellers
- - Locations over the waterline (More than 2 feet)
- - Long distances in between the trough full valve and the raw water pump.
- - Proximity of the exhaust gases pipe with cabin grills (Less than 15 in)
- - Excessive separation between the battery and the Genset

3.- Prepare fittings, hoses, and filters to install the fuel system.
4.- Prepare Hoses, elbows adapters, and Muffler to install the exhaust system (Take into consideration antisiphon devices and Water separators).
5.- Prepare battery cables. Size those cables according to the distance in between the battery and the starter. Always install a Battery switch selector.
6.- Ask the owner his preference to install the remote panel and pass the harness avoiding proximity with Antenna cables, AC cables, and Lightning conductors.
7.- Design a secure frame to support the Generator motor mounts.
8.- Install the sea strainer and raw water pump always below the waterline.
9.- Verify that the emergency generator shutdown system be connected with the fire suppression system.
10.-Setup the AC output power of the generator according to the AC power of the boat (European 230V 50 Hz or American 120/240V 60 Hz).

Chapter 14: Marine Generators

How to install a Marine Generator

11.- Wire the AC output from the generator breaker to the AC switch selector according to the AWG wire gauge.
12.- Connect the generator frame to the boat bonding system.
13.- Start the generator, then apply load starting with light loads finishing with heavy loads always analyzing the generator response (Gain), the voltage, and the frequency.

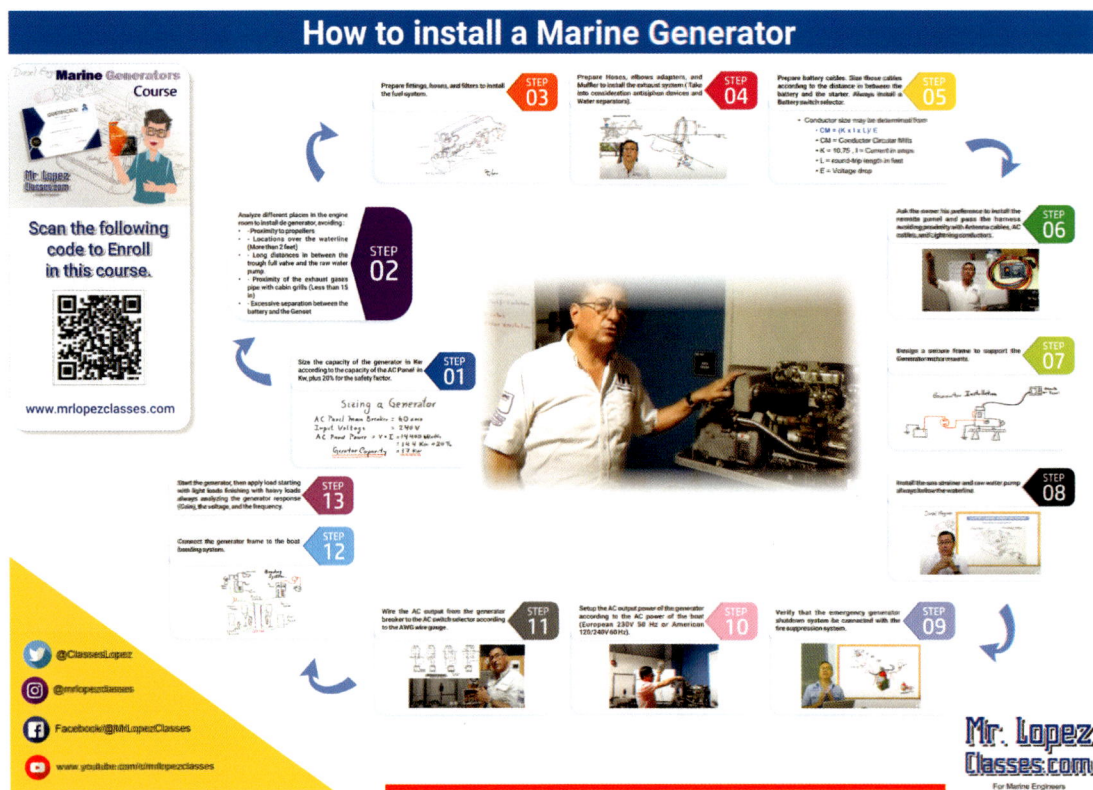

Under Frequency "UF" / "LF"

Low frequency issues are always related with fuel problems or air intake obstructions and consequently with low engine RPM

How to Install a Marine Generator Step by Step

Chapter 14: Marine Generators

Generators Troubleshot
Gasoline & Diesel

Video Episode 4: Troubleshooting Marine Generators

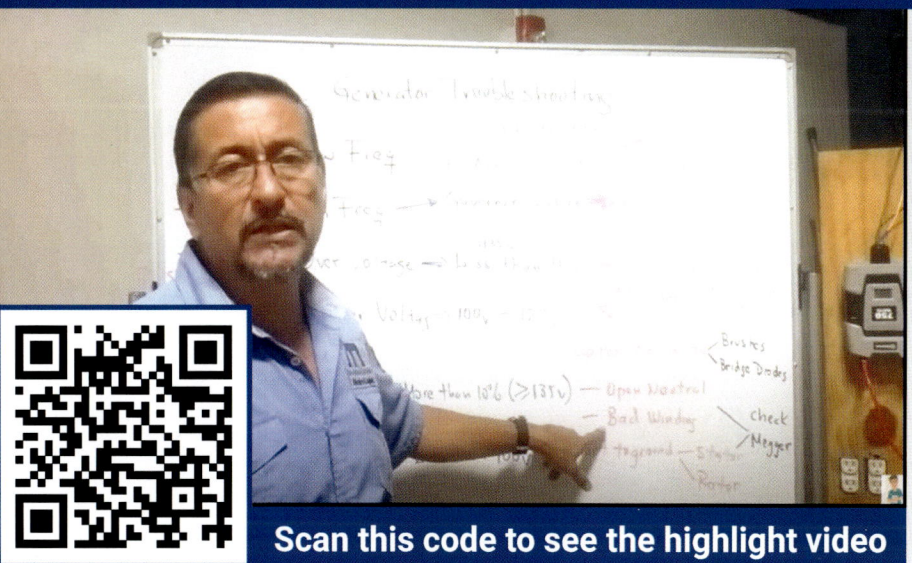

Scan this code to see the highlight video

In this video you will learn the procedure to solve electrical and mechanical problems in marine diesel and gasoline generators analog or digital.
The instructor explain in a simple way how to read faulty codes and how to solve those issues.

Follow me

The Generator won't crank

Primarily this is an electric issue with DC current.
This could be due to a blown 12 volt fuse located on the control panel.
The reading on the voltage meter should be over 10.5 VDC.
A fully charged battery will read 12.6 VCD.
You may possibly have a bad control board or the wiring on the connection may be loose.
No signal on the starter solenoid
A bad Oil Pressure Switch. In general, this is an NC switch.
A bad Coolant Temperature Switch. This is an NC switch.
A bad solenoid valve at the input of the fuel injection pump or
no signal coming in at the solenoid valve.
A blown fuse (DC) on the main control box.
A bad Run Relay at the main circuit board or at the main control box for analog generators.
An open Preheat circuit.
In general, it is a mechanical issue at the engine.

Generator Start but Stop suddenly

Check oil level.
Check oil pressure gauge.
Check oil temperature gauge.
Verify continuity on fuse panel.
Check seal exchanger cap.
Verify coolant level on the heat exchanger and expansion tank.
Check water outlet for the exhaust.
Verify temperature on the raw water pump plate.
Bleed the system manually and check for presence of bubbles on water separator.
Verify if the fuel solenoid is de-energized suddenly (For Diesel engines).
Verify if the Idle solenoid is de-energized suddenly (For Gas carbureted engines).
Verify if the pressure on the common fuel injection rail is according with the manufacturer recommendations.
Verify grounding conditions on the control box.
Check the rotor excitation from the alternator.
Check fuel tank level.
Analyze motor oil conditions by visual inspection.
- Presence of oil
- Presence of coolant
- Presence of metals

Generators Troubleshooting Part II

The Generator Crank but doesn't Start

For Gasoline Engines:
Check for gasoline on the carburetor or fuel injector
Check for spark on the sparkplugs .
Check for presence of moisture on the spark plugs .
Check for presence of oil on the spark plugs.
Verify the free movement on the governor shaft.

For Diesel Engines:

Check for Power on the Glow Plugs.
Bleed the system manually.
Verify Power on Fuel Solenoid.
Check for obstructions on the Turbo Charger.
Verify the Signal on the Fuel Injectors.
Check the Start and Run Relay condition.
Reset the system for more than 30 Sec.
For a Mechanical Fuel Injection System verify the free movement on the governor shaft.
Verify the presence of bubbles on the fuel filters.

Generators Troubleshooting Part III

Low Frequency   (UF)

Fuel Supply.
- Bad filters
- Bad Injectors
- Fuel contaminated
- Air and Water in the line

Fuel Pump.
Fuel Solenoid.

Under Voltage   (UV)

Fuses on Panel-Board.
Control Module.
Open or shorted stator or Rotor.
Low Power on the exciter.
- Bad or contaminated brushes
- Isolated coil on Armature
- Fuel restrictions
- Air restrictions
- Engine loss of power
- High engine temperature
- High exhaust gasses temp.

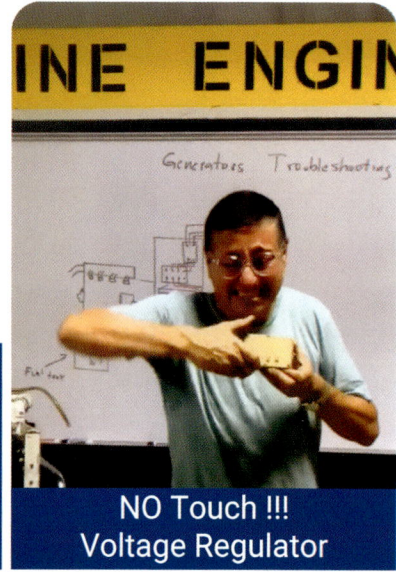

NO Touch !!!
Voltage Regulator

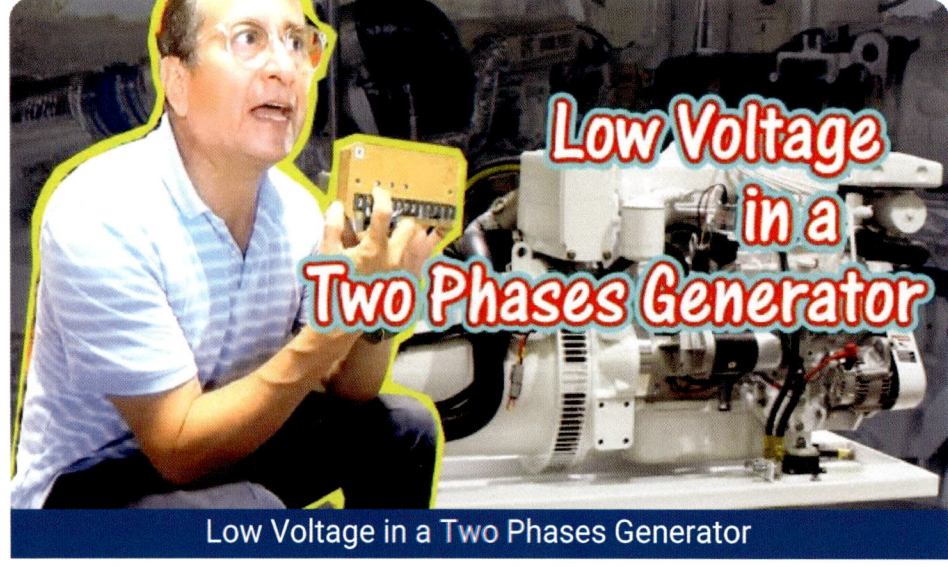

Low Voltage in a Two Phases Generator

Chapter 14: Marine Generators

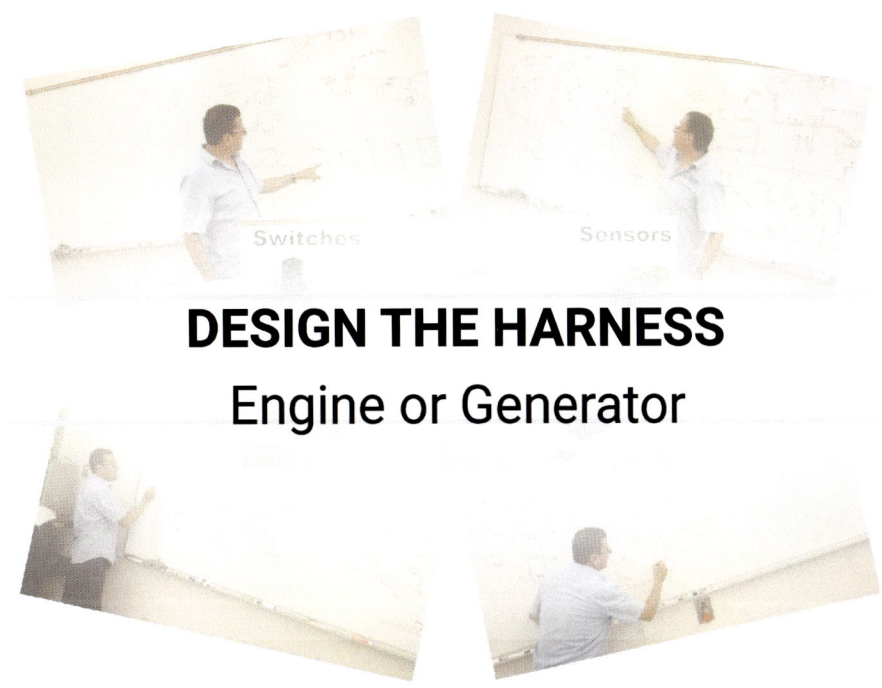

DESIGN THE HARNESS
Engine or Generator

Video Episode 5: How to design the Harness of an Engine or Generator

This video is related with the procedure to build the harness for a Gas or Diesel engine. The harness include the wiring of the engine control box including gauges, switches, relays and fuses and finishing into the sensors and accessories such as alternator, tachometer, starter.

Follow me

Scan this code to see the highlight video

Marine Generators / Oral Test

CHAPTER 15

Inverters and UPS

TOPICS

15.0 Inverters	203
15.1 How Inverters Work	204
15.2 Types of Signal	205
15.3 Input/Output Power	206
15.4 Input Power Vs Output Power	206
15.5 Increasing the Output Capacity	208
15.6 UPS	209
15.7 Sizing the Battery Bank	209
15.8 Example of Inverter Batt. Bank Calculation	210
15.9 Fuses and Breakers AIC Rating	211

Video Episode 1: Inverters – Sizing, Installation and ABYC Standards

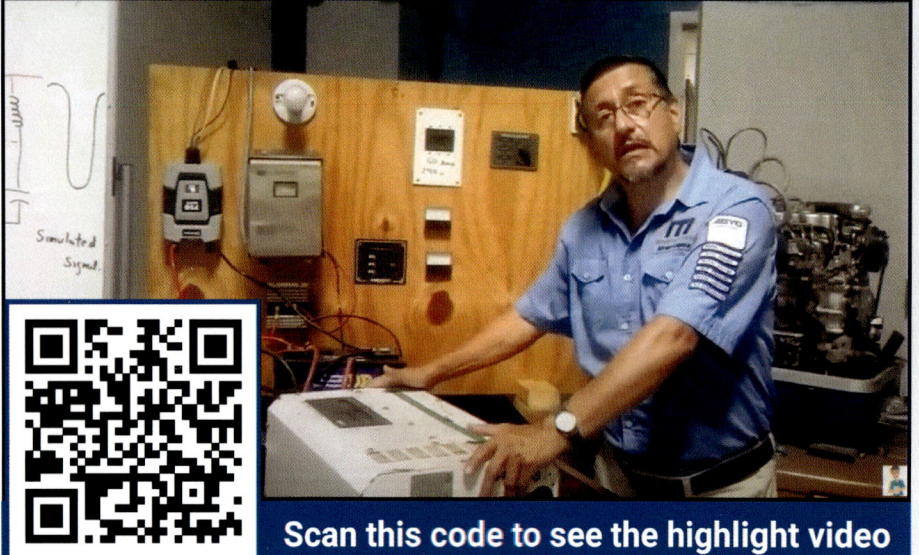

Scan this code to see the highlight video

In this video you will learn how to size an Inverter according with the power of the AC panel in watts. How to calculate the capacity of the battery bank in Amp per hour according with the capacity of the Inverter in watts and how to troubleshoot the most common faults in a typical marine inverter.

Follow me

Chapter 15: Inverters and UPS

Inverters

An inverter is an power electronic device that converts direct current (DC) to alternating current (AC); the converted AC can be at any required voltage and "frequency" with the use of appropriate transformers, switching, and control circuits.

Inverters are usually used to supply AC power from DC sources such as solar panels or batteries. It will be useful for emergency electric source.

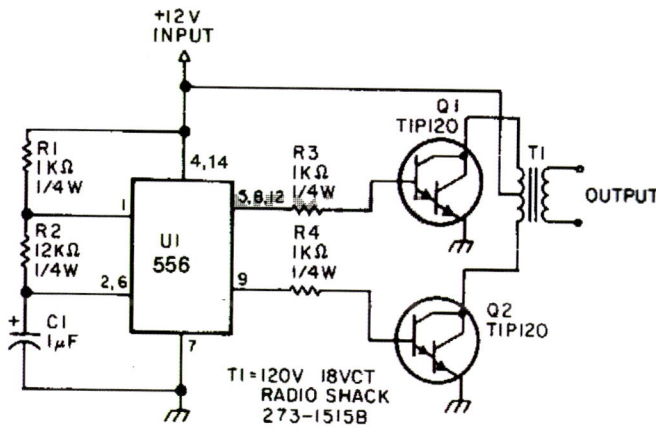

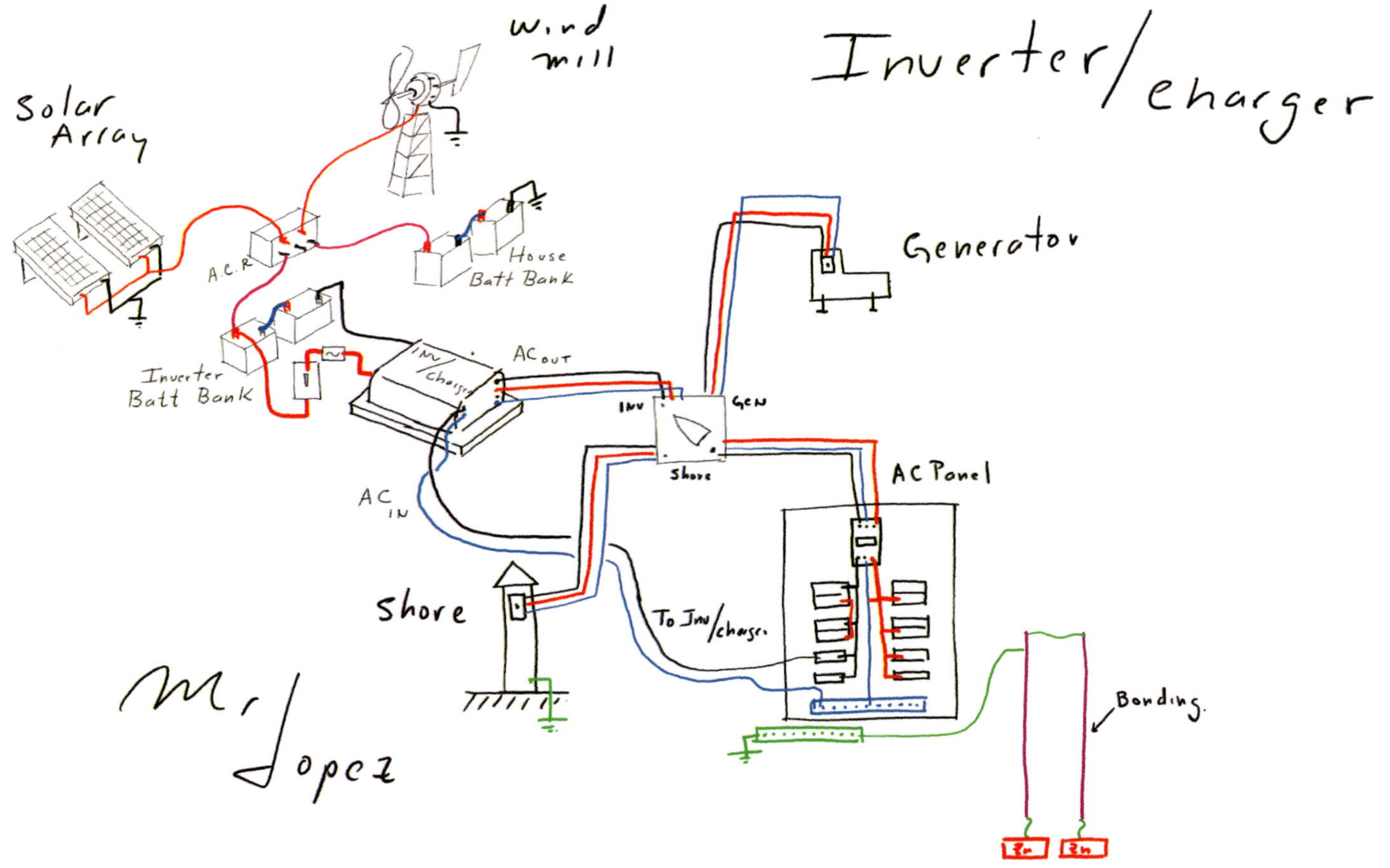

How Inverters Work

DC power, from a hybrid battery for example, is fed to the primary winding in a transformer within the inverter housing. Through an electronic switch (generally a set of semiconductor transistors), the direction of the flow of current is continuously and regular broken (the electrical charge travels into the primary winding, then abruptly reverses and flows back out).
The in/out flow of electricity produces AC current in the transformer's secondary winding circuit.

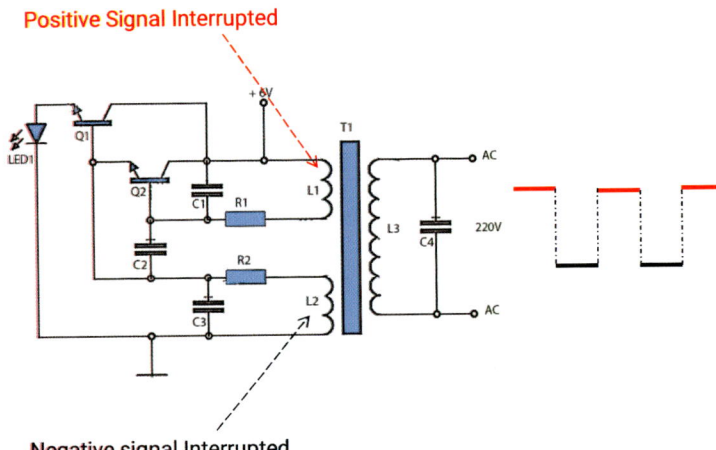

Output Signal

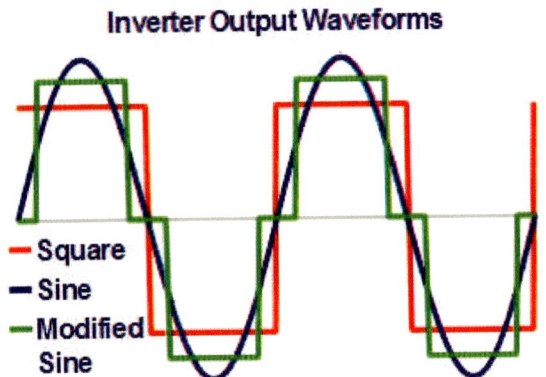

According to output voltage form they could be rectangle, trapezoid or sine shaped. The most expensive, yet at the same time the best quality inverters, output voltage in sine wave.

Inverters fundamentals

Types of Signal

There are three basic types of dc-ac converters depending on their output waveform: **square wave**, modified sine-wave, and pure sine wave (see the diagram).

The **square wave** is the simplest and cheapest type, but nowadays it is practically not used commercially because of low power quality.

Square wave inverters are notorious for causing 60 Hz hum and interference.

To the antenna or input transformer of any electronic equipment, the square wave appears to be a group of sinusoids at multiples of the fundamental 60 Hz.

Whether these higher frequency components interfere depends a great deal on the quality of the equipment's grounding, shielding and power-supply filtering.

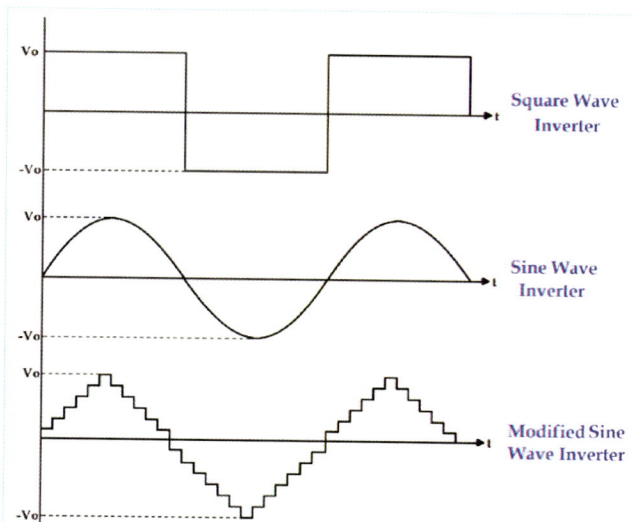

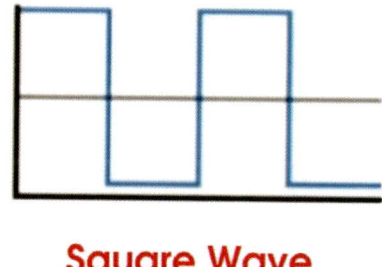

Square Wave

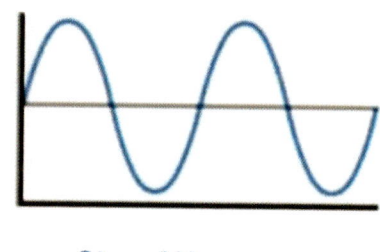

Sine Wave

The **modified sine wave** topologies (which are actually modified squares) provide square pulses with some dead spots between positive and negative half-cycles. They are suitable for many electronic loads.
Priced in the range of $.05-$0.10 per watt, models that employ such a technique are the most popular low-cost inverters on the consumer market today, particularly among car inverters.

High quality pulse-width-modified, sine wave inverters emit less harmonic power but sometimes still cause difficulty in sensitive equipment.

Types of Signal

A true **sine-wave inverter** produces output with the lowest total harmonic distortion (normally below 3%).

It is the most expensive type of AC source, which is used when there is a need for clean sinusoidal output for some sensitive devices such as medical equipment, laser printers, stereos, etc.
For Zero interference and hum, pure sine-wave inverters are available at significantly higher cost.

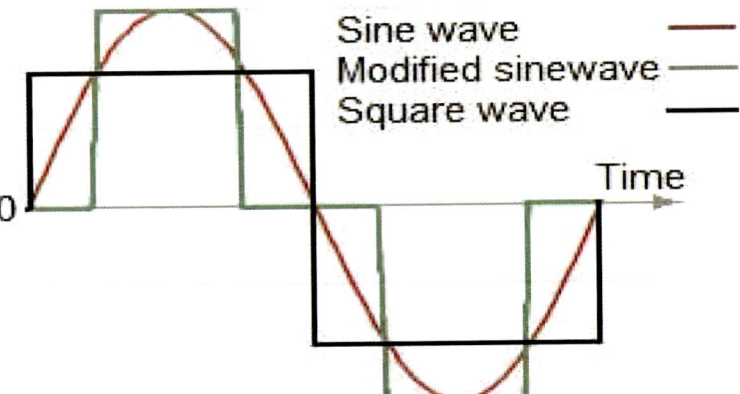

Input Power Vs Output Power

Early inverters for the consumer market were used mainly for mobile applications like boats and recreation vehicles, and most were designed for 12-volt DC battery ignition systems.

Due to an upper capacity limit of approximately 200 amps for the internal power components and heavy welding cables that were being used for connecting typical mobile 12-volt systems, 2,400 watts was about the largest capacity inverter that could be made for these applications (12V x 200A = 2,400 W).

To keep inverter costs low and the designs simple, these early inverters generated a "modified" sine wave output to simulate the 60-cycle 120-volt AC line voltage. The more "steps" in the **modified** waveform, the closer the output voltage will be to a normal AC sine wave.

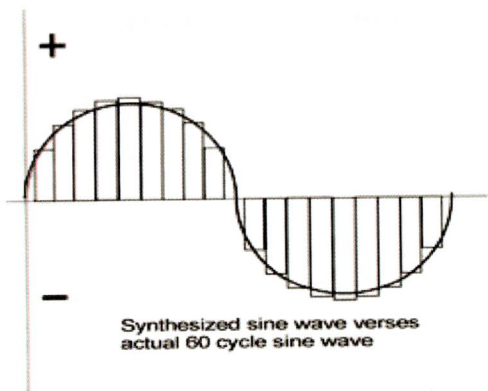

Output AC Performance

Until the explosive increase in personal computers and microprocessor controlled appliances and audio/video equipment, most electrical loads that included older technology would work fairly well on a modified sine wave inverter.
Incandescent lights and power tools also worked well, although some fluorescent fixtures and light dimmers had problems.
An AM radio may produce an objectionable hum in the background, and a microwave oven will take much longer than normal to cook the same food, but most of these devices would still operate on a modified sine wave inverter.

Types of Signal

Power inverters for cars often come with a jack that can be plugged into the cigarette lighter. Note, however, that the cigarette lighters are protected by a fuse rated typically between 15 and 20 A. This is usually enough to run your laptop or other portable electronics.

Output AC Performance

In the early 1990s, quality modified sine wave inverters were being sold by Trace Engineering and Heart Interface. Although still limited to about 2500-watt output using a 12-volt DC input, these became the standard for residential off-grid and back-up power systems.
Manufacturers responded with inverter outputs up to 5,500-watts by using higher voltage 24 to 48-volt DC inputs and more sophisticated internal electronics to increase the number of "steps" simulating the 60-cycle sine wave.

Output Peak Voltage

The outputs of all inverters are pulse-width-modified sine waves, were the duration of the pulses are adjusted to produce a constant rms (root mean square) voltage
The actual peak voltages corresponding to 120 Volt AC rms are:

- Sine wave   170 Volt
- Square wave 120 Volt
- Modified sine wave 150 Volt

Increasing the Output Capacity

Inverter input voltage depends on inverter power, for small power of some 100 W the voltage is 12 or 24 V, and 48 V or even more for higher powers. For large systems 3-phase inverters are available in the market Large inverters could be connected in parallel when higher powers are required.

Uses and Applications

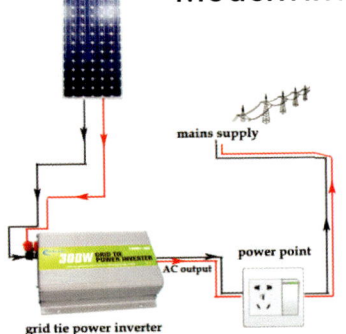

Modern inverters are the most sophisticated electronic devices implemented in photovoltaic systems (See Chapter 16).
On top of high reliable electronics, which must be used, great care should also be taken on lightning protection.

Uninterruptible power supply (UPS)

Is an electronic device that continues to supply electricity to the load for a certain period of time during a utility failure or when the line voltage varies outside the normal limits. Its typical application is PC backup power.

From a technical standpoint, to make a power supply uninterruptible, you need an energy storage backup battery, an AC-DC charger and a DC-AC inverter.

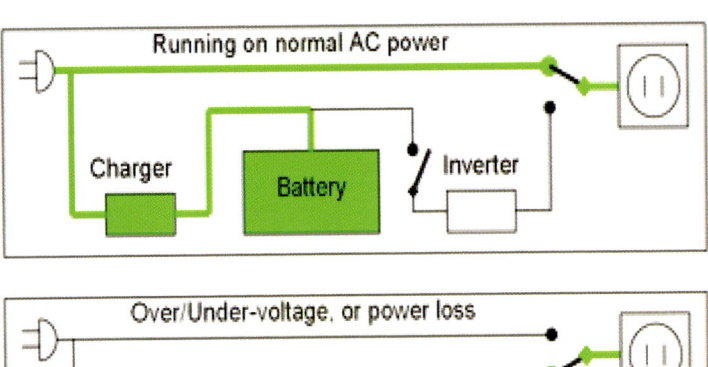

Standby (UPS)

A *Standby UPS* (SPS) includes a transfer relay that switches the load to the battery-powered inverter when the primary AC source fails.

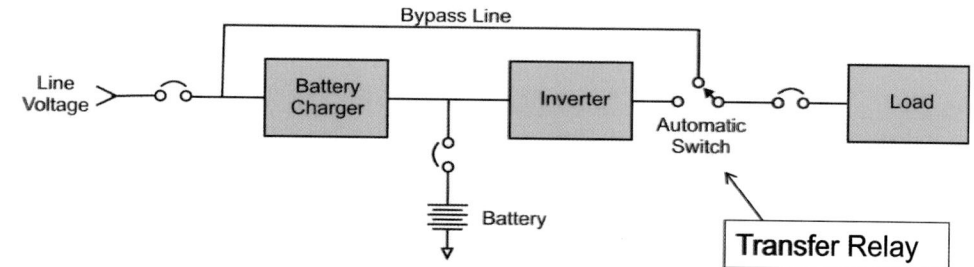

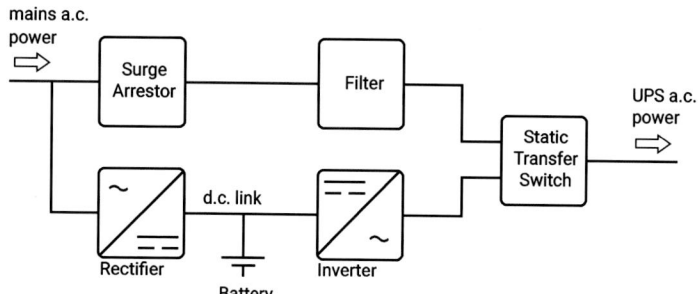

A typical transfer time is between 2 ms and 10 ms depending on the amount of time it takes to detect the lost utility voltage and turn on the DC-AC inverter. During this time the power to the load is momentarily interrupted.

Sizing an Inverter Battery Bank

Watts = Volts x Amps.
Battery capacity is expressed by the amount of Amps per hour that a battery can deliver to support the load applied
For a 12-Volt inverter system, each 100 Watts of the inverter load requires approximately 10 DC Amps from the battery
For a 24-Volt inverter system, each 200 Watts of the inverter load requires approximately 10 DC Amps from the battery.
The first step is to estimate the total Watts of load, and how long the load needs to operate
This can be determined by looking at the input electrical nameplate for each appliance or piece of equipment and adding up the total requirement.
For example, a full-sized refrigerator (750-Watt compressor), running 1/3 of the time would be estimated at 250 Watts-per-hour.
After the load and running time is established, the battery bank size can be calculated.
The first calculation is to divide the load (in Watts) by **10 for a 12-Volt system** or by 20 for a 24-Volt system resulting in the number of Amps required from the battery bank..
The batteries will need to deliver 24 Amps to run the refrigerator (240 Watts/10 Volts = 24Amps/Hr).

Chapter 15: Inverters and UPS

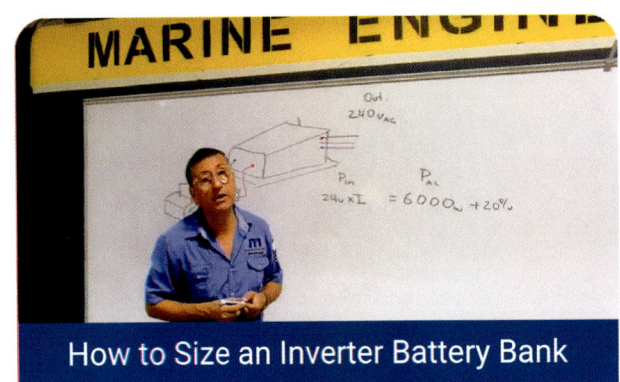
How to Size an Inverter Battery Bank

Example of Inverter Batt. Bank Calculation

Sizing an Inverter Battery Bank
$P_{out} = 6000$ Watts — 240v

DC Side (input) $24V_{DC}$ — INV — AC Side Output $240V_{AC}$

6000 watts + 20% S.F
$P = 7200_w = V \times I$
$I = \frac{7200}{24} = 300$ A/h

6000 watts = $V \times I$
$240v \times I$
$I = 25$ Amp/h

You need a battery bank of 24V and 300 A/h to produce ⟹ 6000 Watts @ 25 A/h During (1) one hour.

Power (in) DC = Power (Out) AC
Watts = $V \times I$ = Watts = $V \times I$

Power (in) + 20% Safety Factor = Power Out

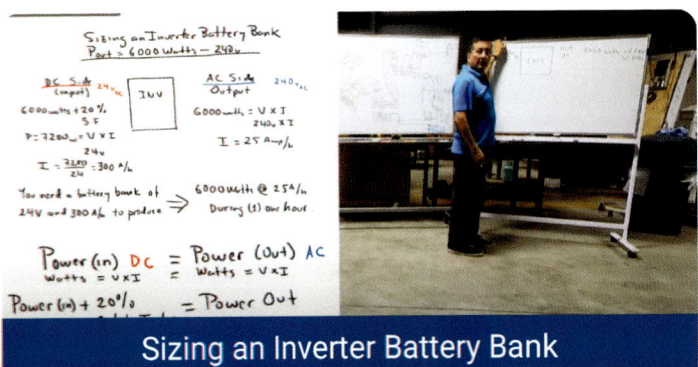
Sizing an Inverter Battery Bank

Fuses and Breakers AIC Rating

According to ABYC, there are some recommendations regarding the protection on the DC side of the Inverter. All fuses and breakers have an AIC rating. The positive cable coming from the battery bank should be protected with a class "T" fuse rated at no more than 20,000 amp AIC. This should not be confused with its nominal rating, around 200 to 300 amps.

AIC Interrupter

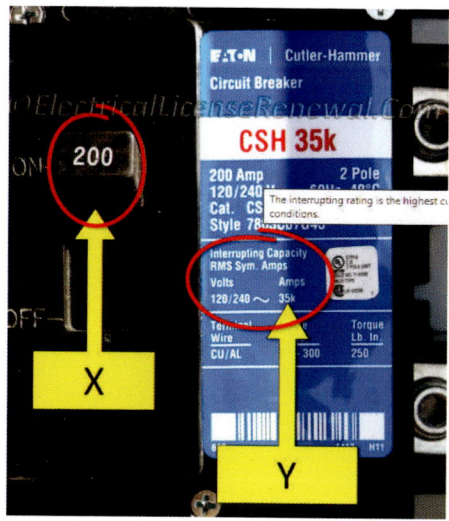

The **interrupting rating** is defined in the NEC as the "**highest**" current at rated voltage that a device is identified to interrupt under standard test conditions.
The value marked on the handle is actually the **ampere rating** of the device (Breaker).
The same 200-ampere rated circuit breaker might also have an interrupting rating or ampere interrupting capacity (AIC) of 35,000 amperes which means that if the breaker is subjected to up to 35,000 amps of current during a fault condition, the device will interrupt the fault condition without blowing up.

A.I.C Rating

AIC Rating is the amount of amperes that a fuse or breaker can be subjected to support without locking or welding itself into the permanent closed position.
For Example a fuse class "T" is rated as 20,000 Amps AIC. However its nominal rating is in between 200 to 300 amp.
The term AIC applies to protective interrupting devices such as circuit breakers and fuses. When a product has an AIC rating, that means it includes circuit protection.
The unit of measure for AIC is Amps RMS Symmetrical. This figure, which is typically in range of 5000 to 200,000 amps, describes the maximum fault current that the protective device can clear safely.

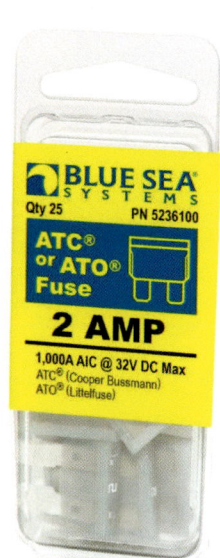

CHAPTER 16

Wiring Diagrams

TOPICS

16.0 Electrical & Electronic Components	213
16.1 Schematics	214
16.2 Wiring Diagrams	214
16.3 Rules	215

Video Episode 1: Wiring Diagrams

Scan this code to see the highlight video

This Video Clip is the faster version of the Episode : "Chapter 14 EP 5 How to design the harness of an Engine or Generator"

Follow me

Chapter 16: Wiring Diagrams

Electrical / Electronic Components

In my book about marine electronics will study in detail all electronic components and their functions in electronic circuits.

Every electrical component, such as a resistor, capacitor and inductor, has a standard symbol used to represent it in a circuit. Each component comes in a variety of packages, which differ in size, shape and sometimes color.

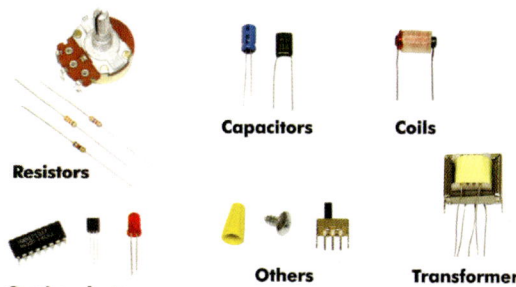

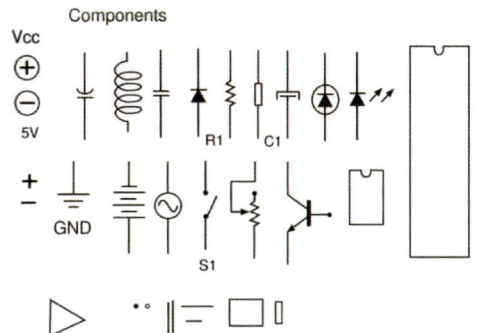

Boat owner manuals and Service manuals have schematics and wiring diagrams to assist in building them and troubleshooting problems.

Wiring Diagram

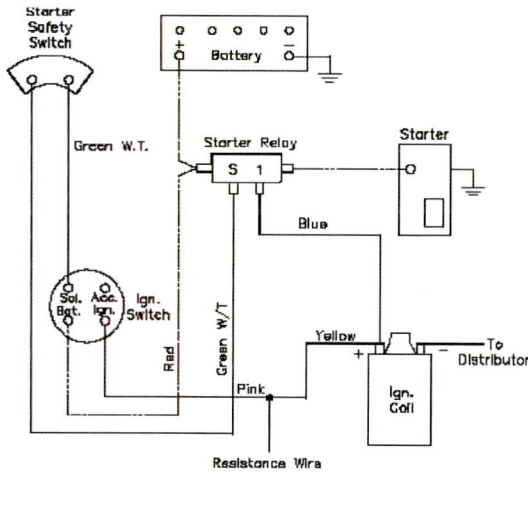

Schematic

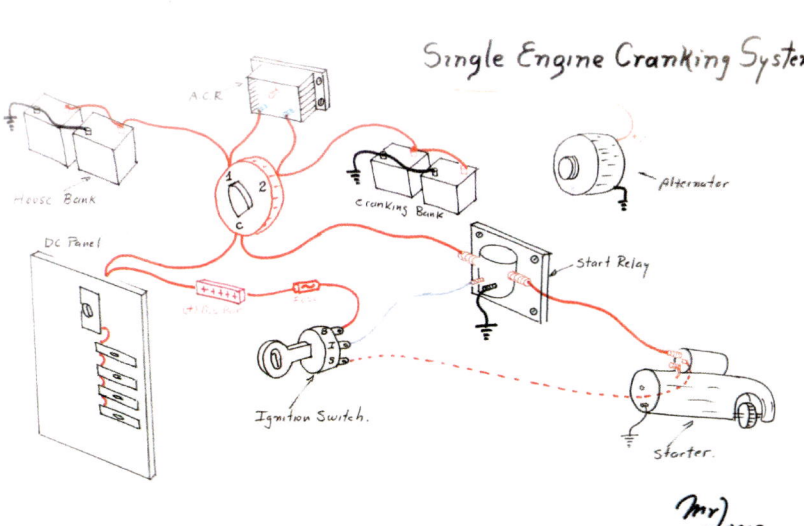

Schematics

The schematics have become very popular because they are visually appealing and easy to follow for a diagnosis.
Schematics are symbolic representations of complete circuits or systems created during the design phase.

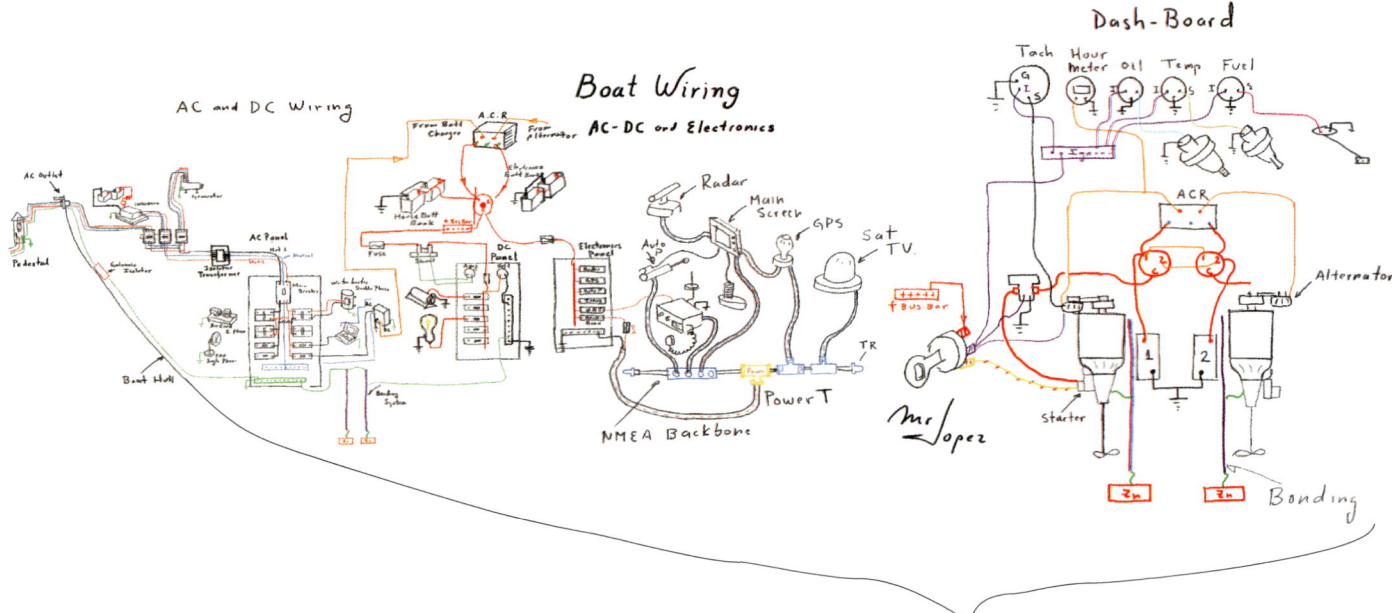

Schematics often assign labels to parts based on their type and arbitrary ordering on the schematic. For example, R stands for resistor, C for capacitors.

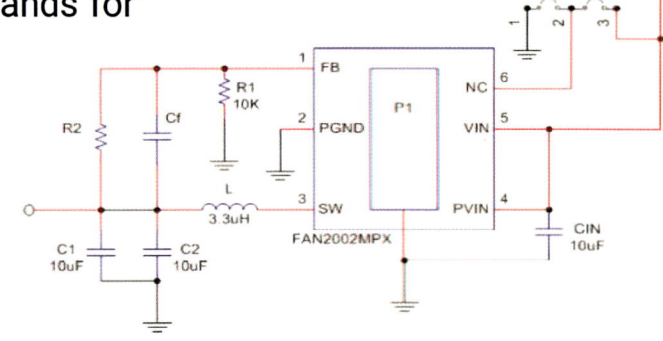

Wiring Diagrams

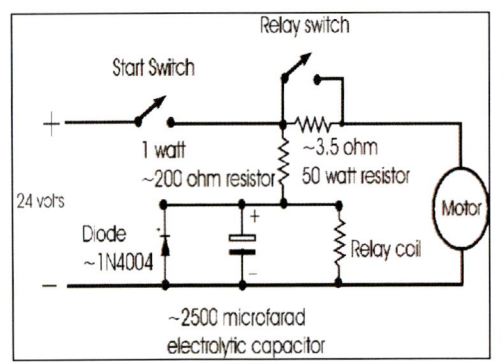

Wiring diagrams, or layouts, illustrate the physical connections, or wiring, between components. They are crucial to the assembly of the circuit or system.
In some cases, the schematic symbol and the wiring diagram symbol are the same. Many wiring diagrams also have a key that provides important information such as wire gauge and colors.

Wiring Diagram Symbols

Schematics often assign labels to parts based on their type and arbitrary ordering on the schematic. For example, R stands for resistor, C for capacitors.

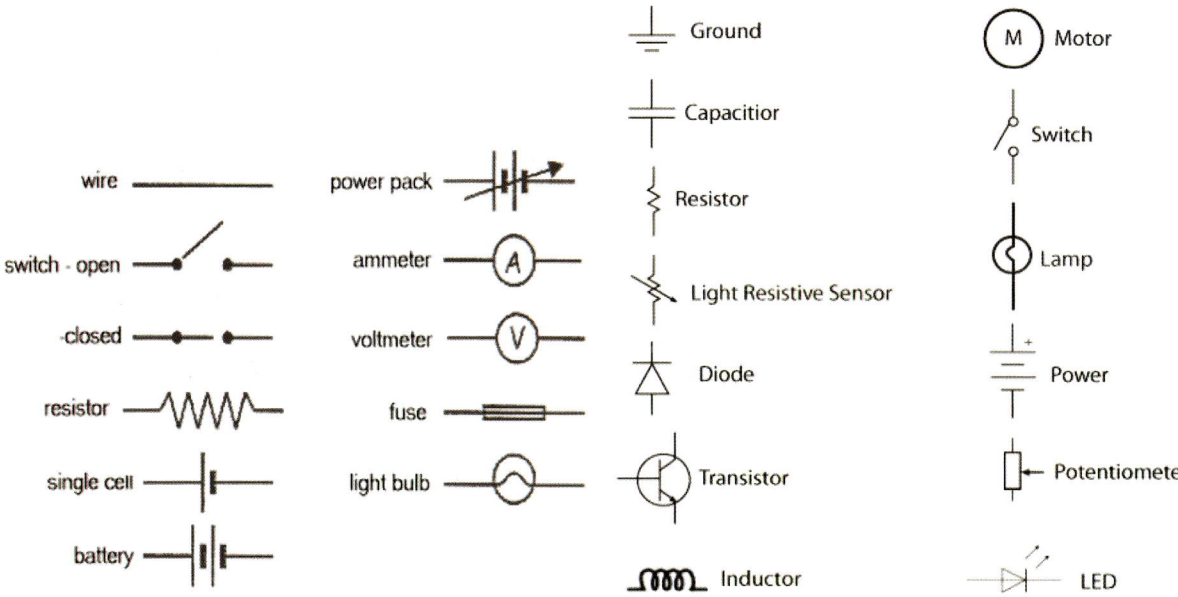

Drawing Schematic Diagram

Schematics and wiring diagrams should be clear and specific. Therefore, part numbers, parts values, polarities, etc., should be clearly labeled to avoid confusion.
A good diagram makes circuit functions clear. Therefore, keep functional areas distinct; don't be afraid to leave blank areas on the page, and don't try to fill the page.

Rules

Wires connecting are indicated by a heavy black dot; wires crossing, but not connecting, have no dot (don't use a little half-circular ``jog"; it went out in the 1950s).
Four wires must not connect at a point; i.e., wires must not cross and connect.
Always use the same symbol for the same device; e.g., don't draw resistors in two different ways.

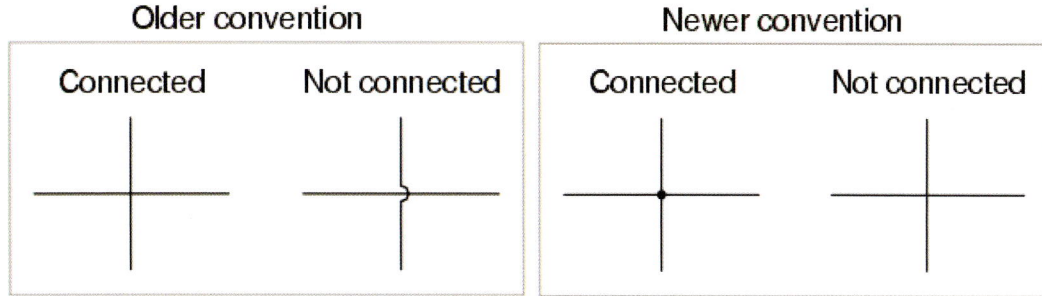

Chapter 16: Wiring Diagrams

Rules

Wires and components are aligned horizontally or vertically, unless there's a good reason to do otherwise.

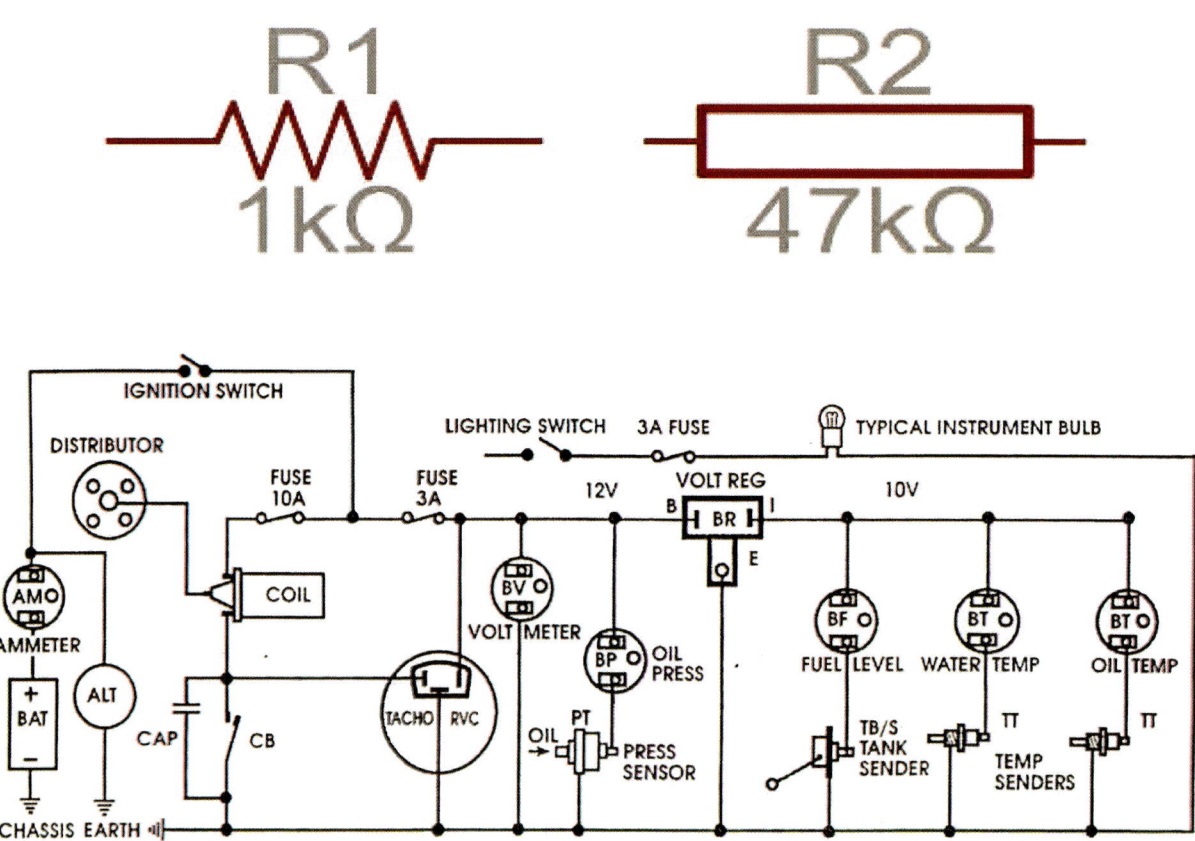

At the end of any Electrical or Electronic installation, a good marine technician should provide detailed wiring diagrams and schematics related to the design of the project, Keep in your files a copy of those designs, in the future those diagrams could be the entry key of a new project.

CHAPTER 17

Solar Power

TOPICS

17.0 The Solar Cell	218
17.1 Photovoltaic Cell	219
17.2 Solar Panel Configuration	221
17.3 The Charger Controller	222
17.4 Solar Panel System Calculation	224
17.5 Boat Solar Panel System Configuration	226

Video Episode 1: Solar Panels Configuration

Scan this code to see the highlight video

In this you will find a valuable information related with the configuration of solar panels for pleasure yachts and sail boats.
In the video Mr Lopez explain in a simple way how to calculate the capacity of the battery bank in Amp per Hour according with the capacity of the panels in Watts.

Follow me

Chapter 17: Solar Power

The Solar Sailor

Can run on wind, sun, battery, or diesel, or in any combination.

Solar power

Solar power is the result of converting sunlight into electricity.

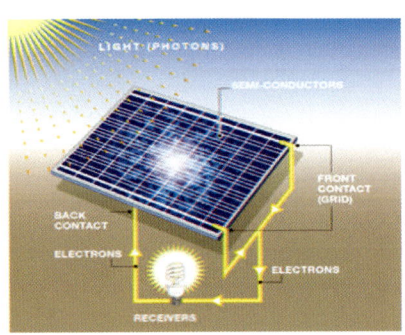

Sunlight can be converted directly into electricity using **Photovoltaic cells** (PV), or indirectly with **Concentrating Solar Power** (CSP), which normally focuses the sun's energy to boil water which is then used to provide power.

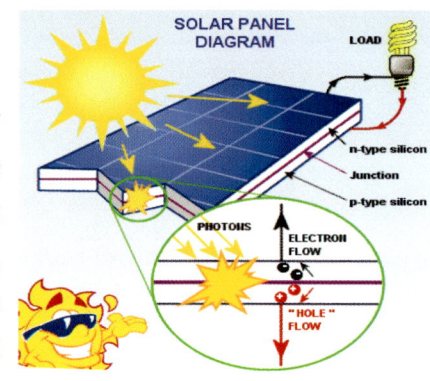

Marine Solar System

With solar power, your marine life can be truly independent.
Solar power can keep your galley appliances, communications, navigation gear, running lights and pumps all running happily.
It is a proven fact that sunlight contains around 1,000 watts of power per square meter.

The Solar Cell

When sunlight hits a doped semiconductor, such as silicon, the photons knock electrons free to produce electricity from their movement.

By putting wires into the solar transducer to draw off the juice, you've created a photovoltaic cell.

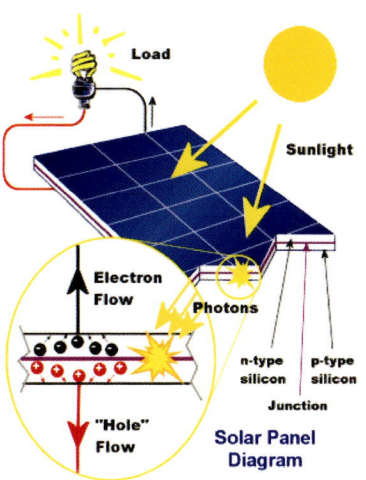

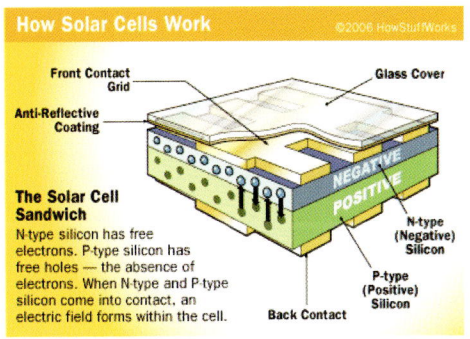

Solar cells are made of the same kinds of semiconductor materials, such as silicon and Germanium used in the microelectronics industry.

Photovoltaics

Photovoltaics is the direct conversion of light into electricity at the atomic level.

Some materials exhibit a property known as the photoelectric effect that causes them to absorb photons of light and release electrons. When these free electrons are captured, an electric current results that can be used as electricity.

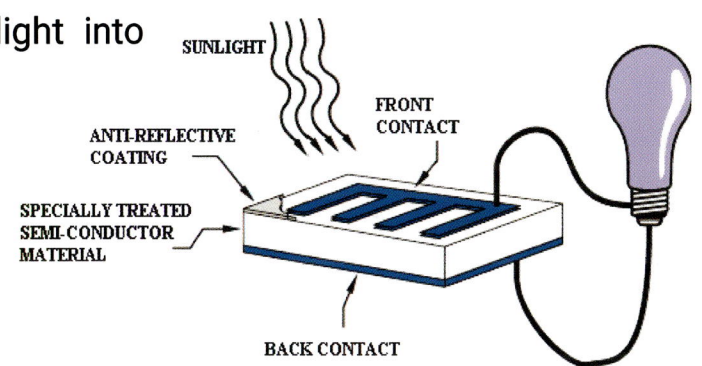

Photovoltaics (PV)

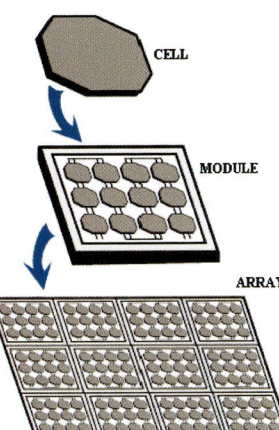

Photovoltaics were initially used to power small and medium-sized applications, from the calculator powered by a single solar cell to off-grid homes powered by a Photovoltaic array.

Chapter 17: Solar Power

What is a Photovoltaic Cell?

A photovoltaic cell is made up of a nonconductor, with one side coated with metal atoms that produce electrons when they are exposed to the sun.

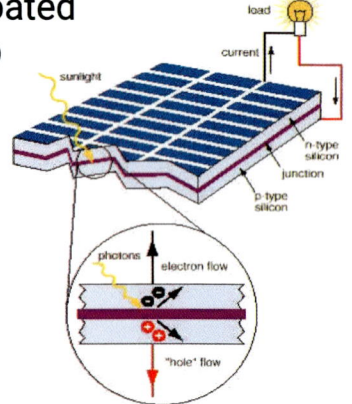

The other side of the PV cell is coated with negative electron atoms. If you connect a wires, from each side of the cell, to a device, current will flow when the positive side is exposed to sunlight.

The Basics of a Solar Electric System

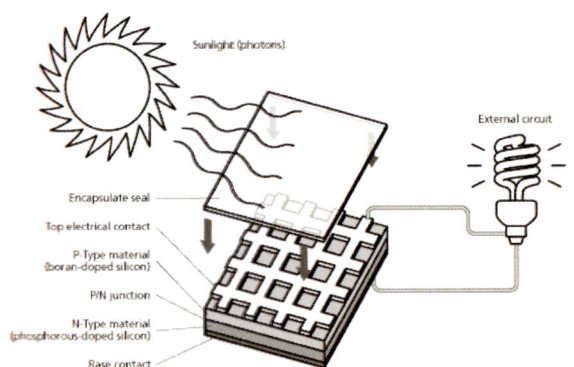

PV technology converts sunlight directly into electricity. Electrons are excited by particles of light and driven toward the surface of the PV cell by an electric field inherent in the semiconductor material of the PV cell.

Solar Cells

For solar cells, a thin semiconductor wafer is specially treated to form an electric field, positive on one side and negative on the other.

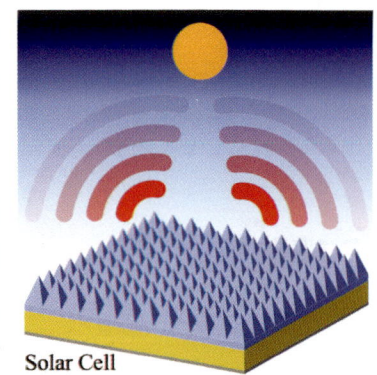

When light energy strikes the solar cell, electrons are knocked loose from the atoms in the semiconductor material. If electrical conductors are attached to the positive and negative sides, forming an electrical circuit, the electrons can be captured in the form of an electric current.

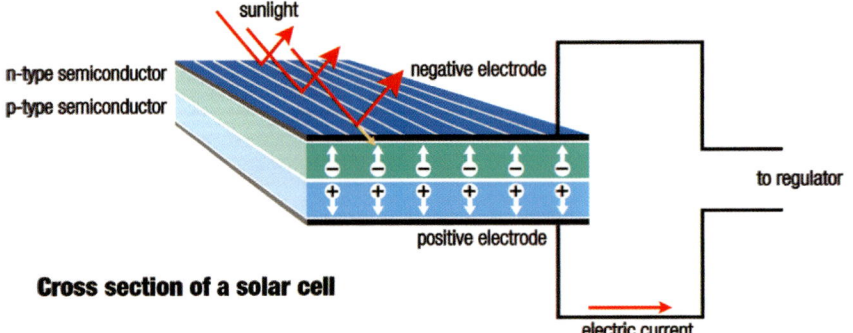

Cross section of a solar cell

The downside of a Photovoltaic Cell

It can be less efficient than other types of power. A small solar panel can only power equipment which doesn't require a lot of power. What's more, when you're outdoors, you can't always expect the weather to cooperate with you.
Chances are, you'll experience bad days when the sun barely peeks out of the clouds. At times like these you're solar-powered gadgets will not work.

Solar Panel

A number of solar cells electrically connected to each other and mounted in a support structure or frame is called a photovoltaic module.
Multiple modules can be wired together to form an array.

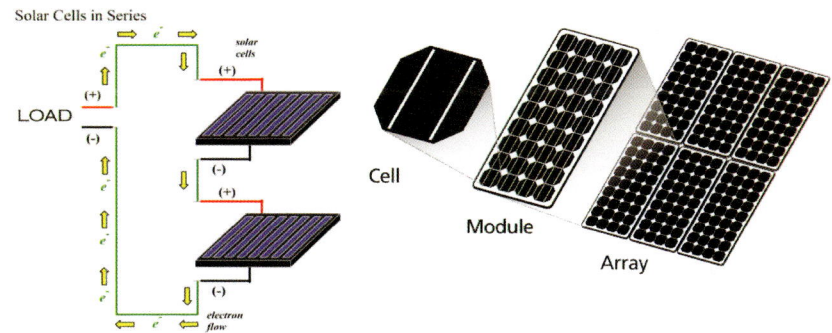

Series and Parallel Arrays

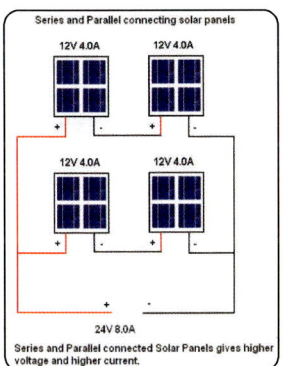

As with the batteries, the solar panels can be connected in series or parallel to increase the output voltage or the capacity.

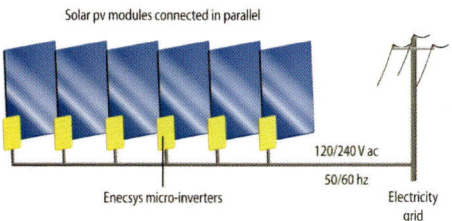

Charging the House Batteries

Remember that the capacity of its solar panel must exceed by 30% the capacity of the battery bank you want to recharge.
You should calculate the capacity of the battery bank depending on whether the batteries are connected in series or parallel.

Series = More Volts VT = V1 + V2 + V3 + ...
Parallel = More Amps IT = I1 + I2 + I3 + ...

Chapter 17: Solar Power

Uses and Applications

Commonly boat owners say the installed solar panel is useless.
The reason is that the installed Solar Panel has a lower capacity than the battery bank.
So the batteries will be fully charged in a long period of time.
A solar panel Array with a capacity in Amp per hour, equal to the capacity of the battery bank or no less than 20% of his total capacity is my recommendation, to keep the battery bank fully charged all the time.

400 Watts, 12 Volts, 140 AH

Charger Controller

You can use the solar panel to charge two battery banks simultaneously, for this you may use a "battery stabilizer" or "Charger Controller" as we saw in the Chapter 6 (Pag.90) of this book.

The charger controller could be installed between the solar panel and the battery bank.

This is to ensure that the batteries don't get over-charged.
It also prevents an electrical current from running back from the
battery to the solar panel at night when no power is being generated by the solar panel.

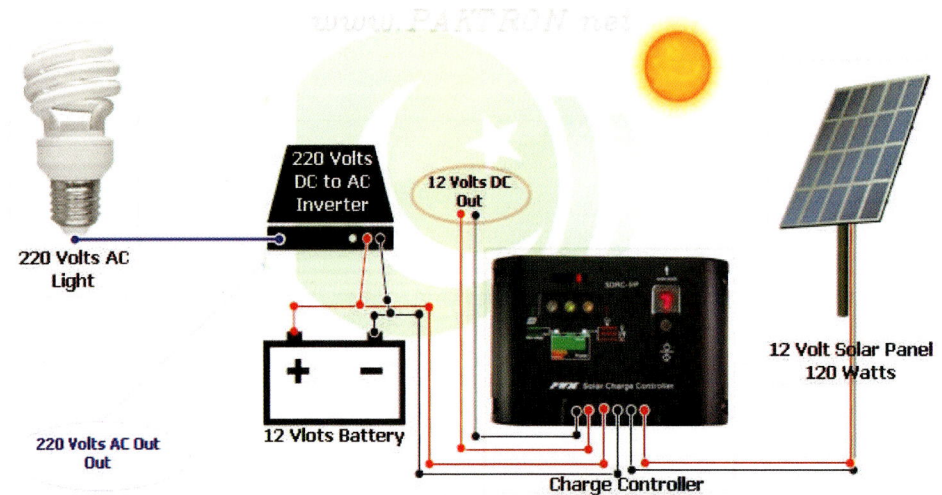

Chapter 17: Solar Power

Charger Controller

Outside View Of The Controller

- Solar Battery Indicator Light (1)
- Storage Battery Indicator Light (2)
- Load Indicator Light (3)
- Work Mode Digital LED Display
- Setting KEY
- + − Solar Battery
- + − Storage Battery
- + − LOAD

Controller and Remote Display

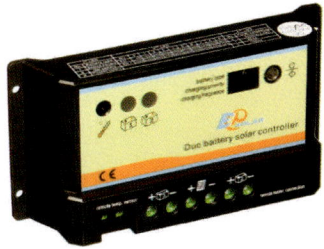
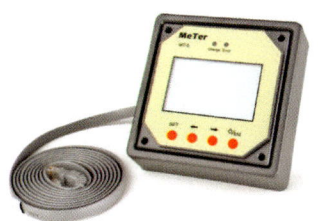

The remote panel helps monitor the system. The indicator lights alert you when the battery charge is low and also when the intensity of the sunlight is low.

Uses and Applications

Never try to power the appliances directly from the solar panel. Both motor efficiency as the solar panel efficiency will be affected.

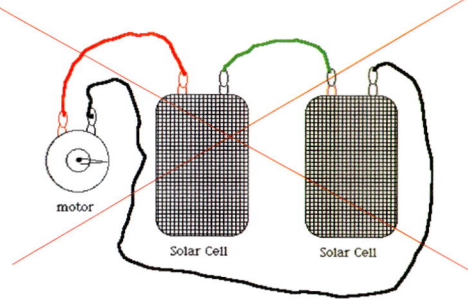

Solar Panel System Calculation

The DC current produced by the Solar Panel can be used in two different ways:
First, it can be used to recharge the batteries that power de DC panel. In that case, it is important to know what the total capacity of the battery bank is
(Amp / Hr). Also can be used to recharge the inverter battery bank. In that case, you need to know a list of all your appliances and devices in the AC Panel and how many hours per day each will be run.
In other words, you need to know the total consumption in Amps per Hour per day in the AC panel.

Solar Panel System Calculation

It is important that you understand that the Inverter is an electronic device that converts DC power into AC power. In other words, if I want for example 6,000 watts of AC power at the output of the inverter I need minimum of 6,000 watts of DC power at the input plus 20% additional due to the dissipation of energy by heat, voltage drop, and other external factors.
In this way, if you want 6.0 Kw at the output you should have at least 6.0Kw + 20% = 6.0x 1.2= 7.2 Kw at the input in the DC side of the system.
Now we are going to calculate how many amps per hour are required to get 7200 Watts of DC power (For a 24 V DC System)
7200 Watts = V x I = 24V x I. Then I = 7200 / 24 = 300 A/H
300 A/H is the minimum capacity recommended for the battery bank to supply 6000 Watts of AC power for one hour.

$$\text{Inverter}$$

$$\underline{\text{DC Side}} \qquad \underline{\text{AC Side}}$$

$$\text{Power (In)} = V \times I = 24_v \times I \qquad \text{Power} = V \times I$$
$$(P_{out}) = 6000 \text{ Watt}$$

$$\text{Power (IN)} = P(out) = 6.0 \text{ Kw} = 240v \times 25A$$

$$\text{Power (in)} + 20\% = 6.0 \times 1.2 = 7.2 \text{ kw} = V_{DC} \times I_{DC} \Rightarrow I_{DC} = \frac{7200}{24} = 300 \text{ A/h}$$

How to Calculate a S.P System

You need to know how much energy your battery can store and then select a Solar panel that can replenish your 'stock' of energy in the battery in line with your pattern of use.
Keep in mind that a good solar panel Array should have a capacity in Amp per hour, equal to the capacity of the battery bank or no less than 20% of its total capacity, to keep the battery bank fully charged in a short period of time.

Keep in mind that to protect the life of the battery it is recommended never discharge it below 50%.

Chapter 17: Solar Power

Solar Power System

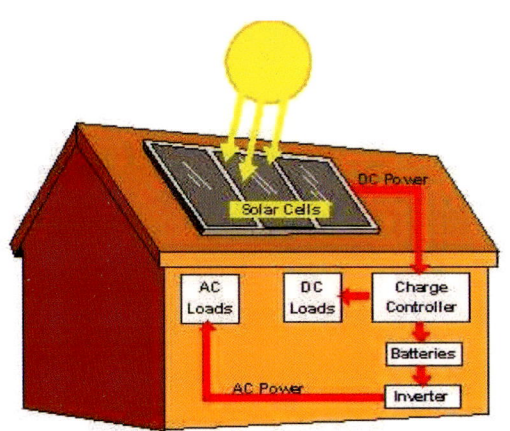

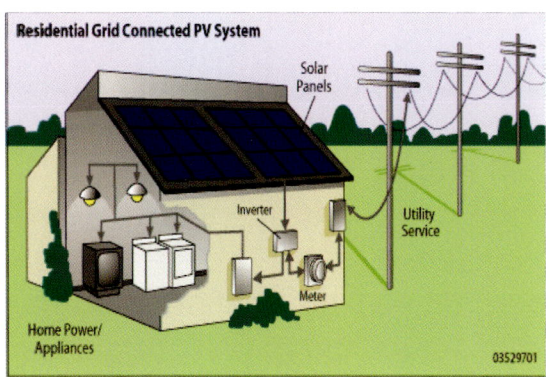

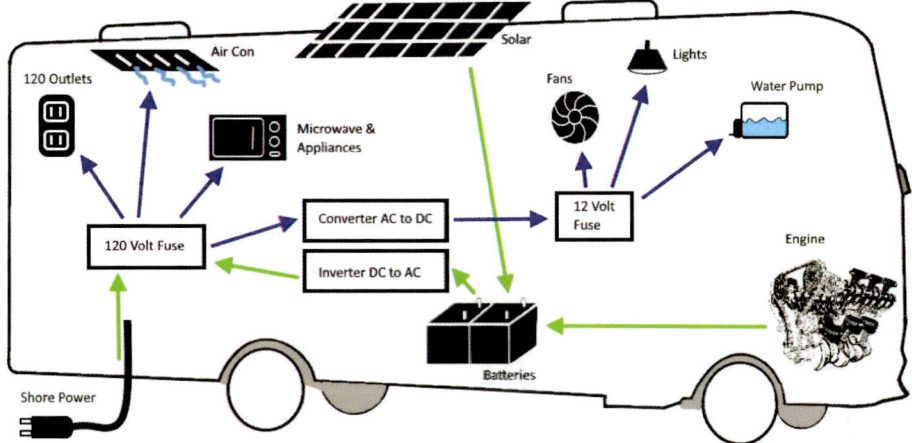

Boat Solar Kit

RV & Boat Solar Kit - 110 Watt Remote Power System.

Reg Price: $1,400.00
On Sale For: $1,285.89

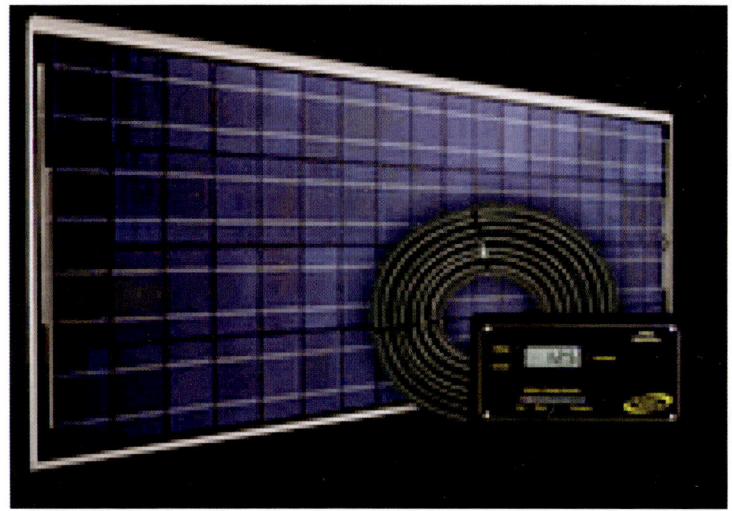

Solar-power kits for boats

Each Solar Marine Power Kit includes a solar module, a charge controller, mounting hardware, heavy-duty wiring, and an installation guide.
The kits are available in a variety of power outputs (50, 75,100 and 400 watts) to meet the various needs of boat owners operating direct current (DC) appliances.
Remember that you can configure arrangements of panels in series and parallel to get the power that you need.

Boat Solar Power System Configuration

CHAPTER 18

Wind Power

TOPICS

18.0 Wind Mill Configuration — 228
18.1 AC / DC Wind Mills — 229
18.2 Types of Wind Mills — 229
18.3 How the Wind Mill works — 230

Solar-power kits for boats

Wind Mill or Wind Generator

A wind turbine is a machine that converts the kinetic energy in wind into mechanical energy.

If the mechanical energy is used directly by machinery, such as a pump or grinding stones, the machine is usually called a windmill.

If the mechanical energy is converted to electricity, the machine is called a wind generator, or more commonly a wind turbine.

Due that the air stream is not constant during the day ,the majority of the wind turbines produce DC current that with a simple voltage regulator can be stored in a battery bank.

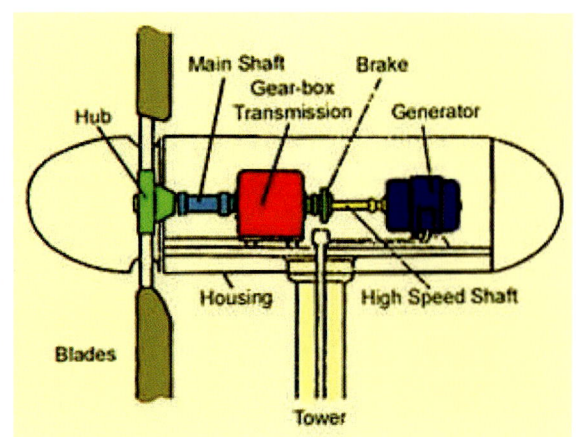

Wind Turbines

They can be used alone, or they may be used as part of a hybrid system, in which their output is combined with that of solar panels and /or an onboard generator.

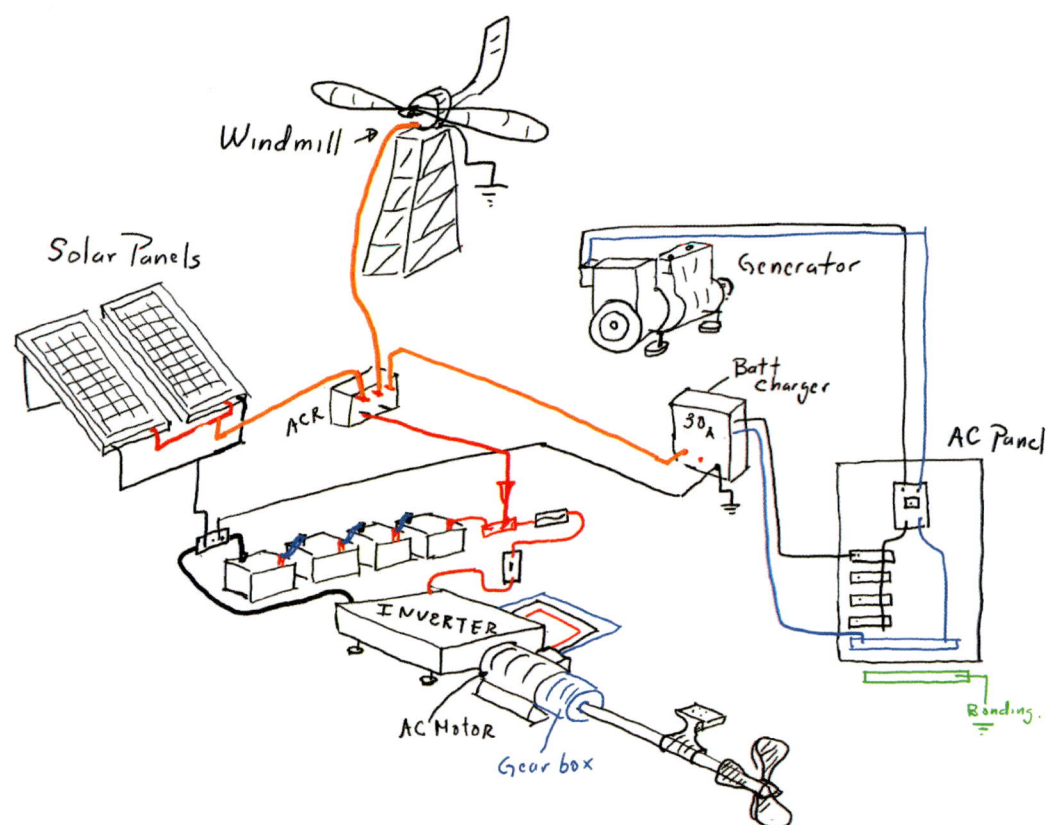

Chapter 18: Wind Power

AC Wind Generator Components

In regions with a constant airstream; large windmills with longer blades and higher gear ratios are designed to produce AC power that is transported to long distances by means of step-up transformers.

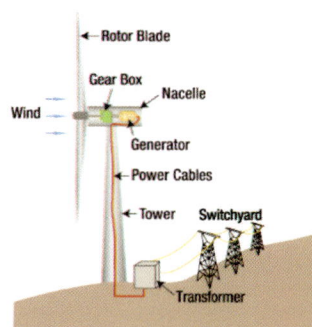
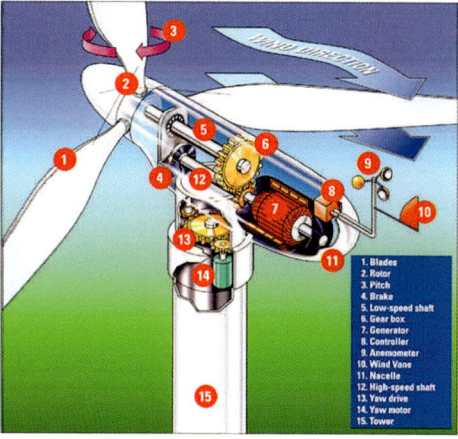

Wind Generators Types

Wind turbines can be separated into two types based on the axis about which the turbine rotates. Turbines that rotate around a horizontal axis are more common. Vertical-axis turbines are less frequently used.

Multi Directional Wind Turbine

Is ideal for urban environments, low draw and off-grid applications. It's inexpensive, reliable and simple, works well in low speeds.

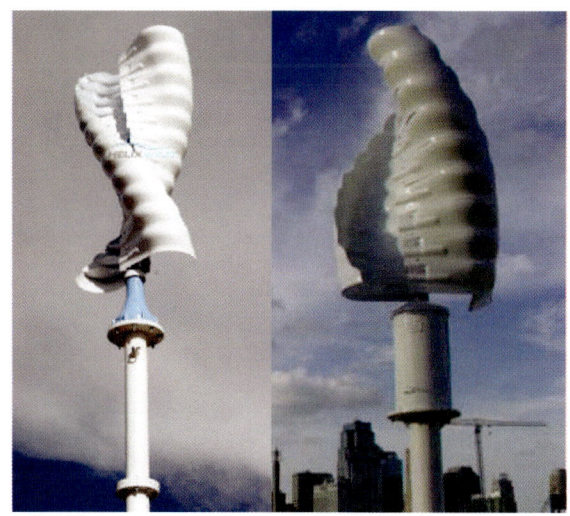
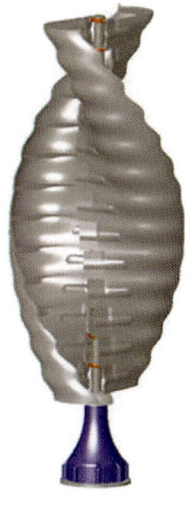

Chapter 18: Wind Power

Wind turbine / Windmill

A **wind turbine** is a rotating machine which converts the kinetic energy of wind into mechanical energy.
If the mechanical energy is used directly by machinery, such as a pump or grinding stones, the machine is usually called a windmill.
If the mechanical energy is instead converted to electricity, the machine is called a **wind generator.**

How the Windmill Works?

Like the solar panel the majority of the windmills produce DC current. This current pass through the charger controller and then is connected with the battery bank.

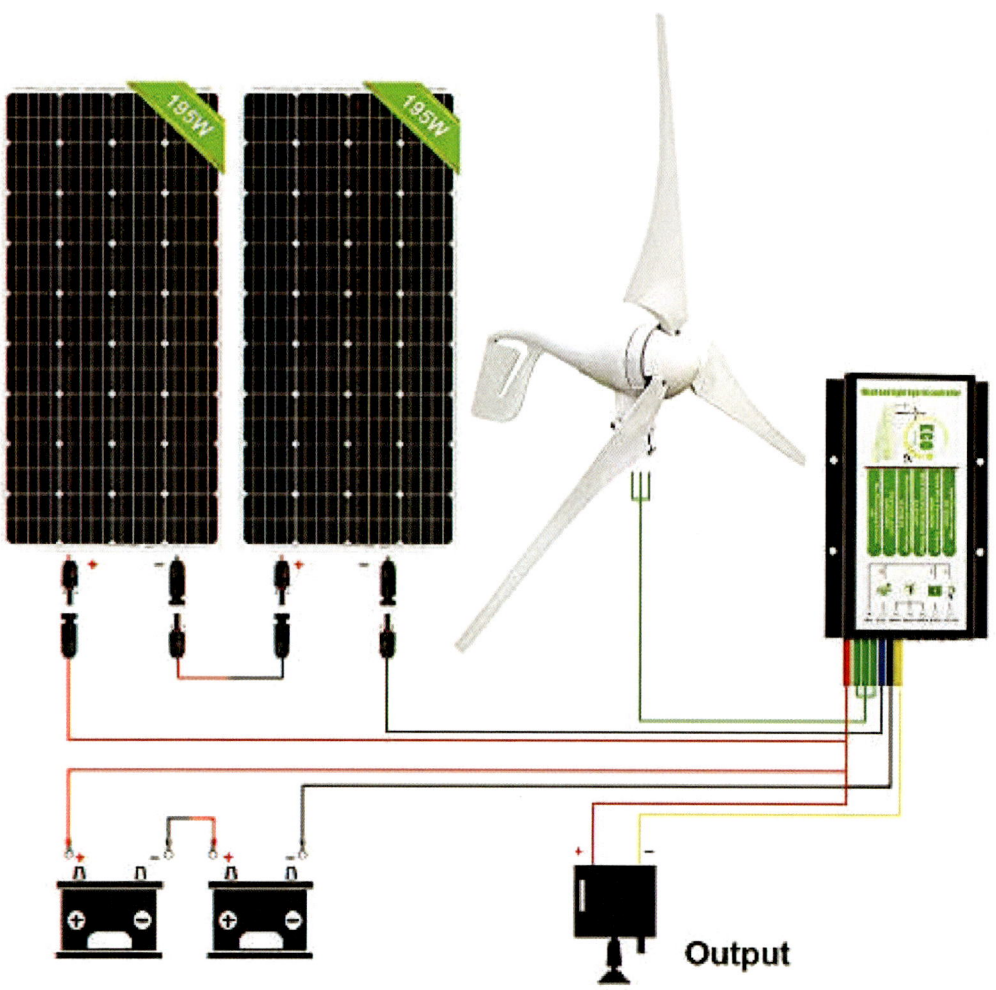

CHAPTER 19

Hybrid Vehicles

TOPICS

19.0 Hybrid and Electric Vehicles	231
19.1 Hybrid Boats	232
19.2 Electric & Hybrid	233
19.3 Parallel Hybrid	235

Hybrid and Electric Vehicles General Features

In the following videos, you can see the detailed process to convert a vehicle (Boat, Car, or plane) into an Electric or Hybrid vehicle.
This is part of my new book. Electric and Hybrid Vehicles.

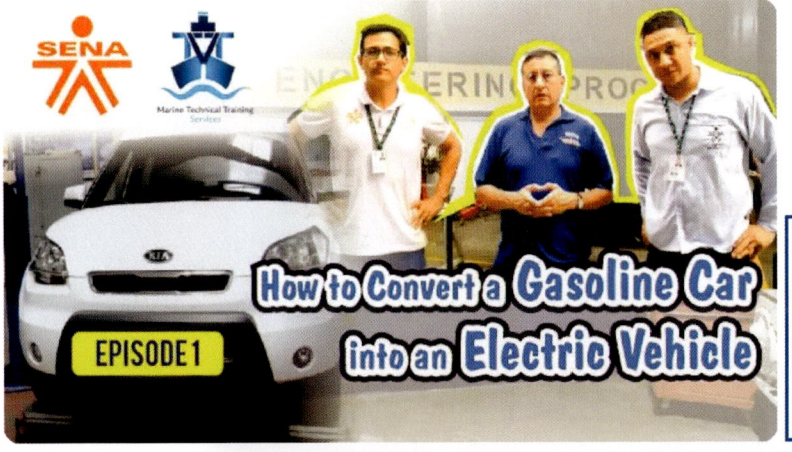

Hybrid and Electric Vehicles General Features

Hybrid Boats

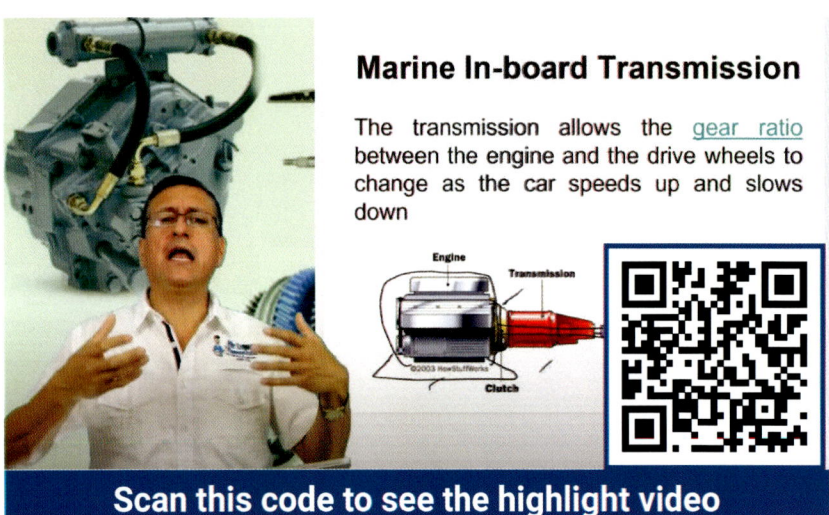

Marine In-board Transmission

The transmission allows the gear ratio between the engine and the drive wheels to change as the car speeds up and slows down

Scan this code to see the highlight video

Video Episode 1
Hybrid Boats Fundamentals Part I

In this video you will learn the procedure to select the appropriate electric motor to replace the original internal combustion engine in your boat. This video is a great tool to understand the meaning of a Hybrid boat.

Follow me

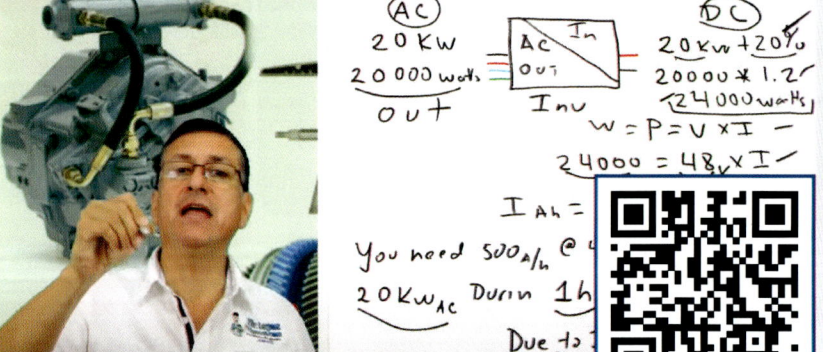

Scan this code to see the highlight video

Video Episode 2
Hybrid Boats Fundamentals Part II

In this Video you will learn the procedure to size the battery bank and the solar panel array according with the electric motor selected to replace the original internal combustion engine in your boat.

Follow me

Hybrid Boats

With the new configurations, you can run your boat in different ways: Fully electric AC or DC (Depending 100% on Solar Panels and Windmill) and Hybrid, Fully electric AC or DC with intermittent support of a Generator (Just to recharge the Battery Bank).

The propulsion in a boat can be achieved:
By means of an internal combustion engine fuelled with , Diesel, Gasoline or LPG. that transfer the output torque directly to the propeller.
Using the force of the wind against a sail
Using an AC motor powered from a Generator or an Inverter. There are three hybrid configurations

- – Diesel / Electric
- – Serial Hybrid
- – Parallel Hybrid

Hybrid Boat (Diesel/Electric)

The power produced by a generator feed an AC motor, and the output torque is transferred to the propeller shaft via a motor controller.

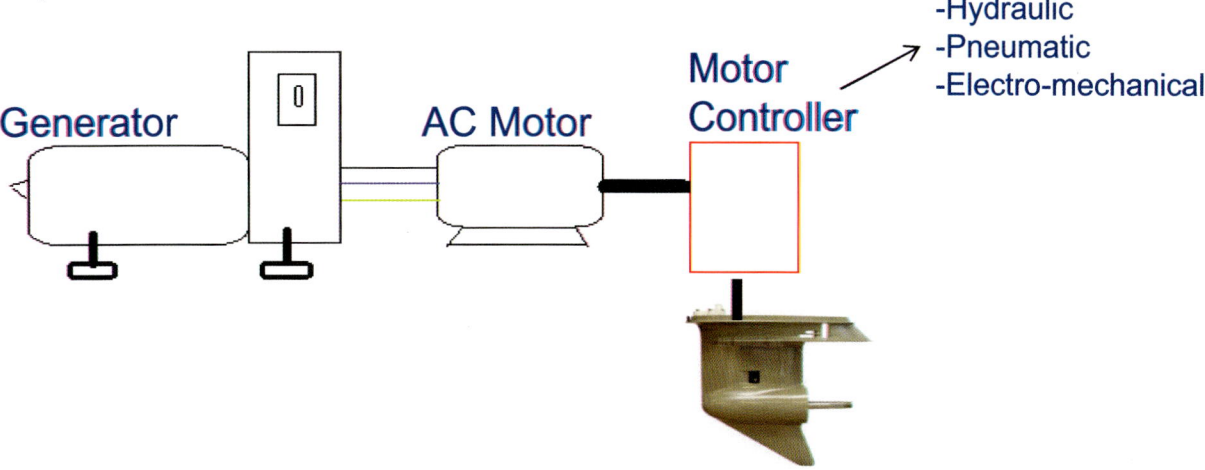

Hybrid Boat (Serial Hybrid)

In this option the system is powered entirely by DC power. The generator produce DC current that is sent in two ways through the charge controller, to keep the charge in the bank of batteries, and to power the DC motor.

When the generator is no running, the energy storage in the battery bank is used to power the system.

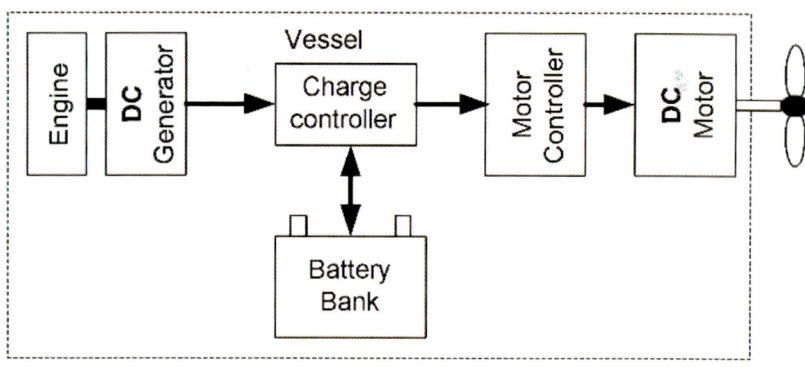

Electric & Hybrid

New engine Compartment.
Pedestal with a rectifier from 120 AC into 48 V DC.

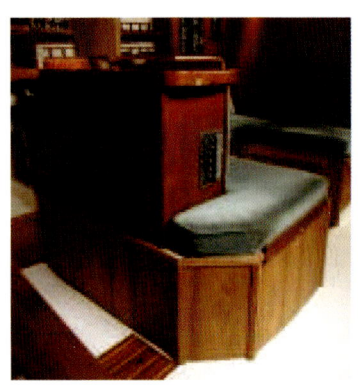

Engine Compartment (Brushless DC Motor).

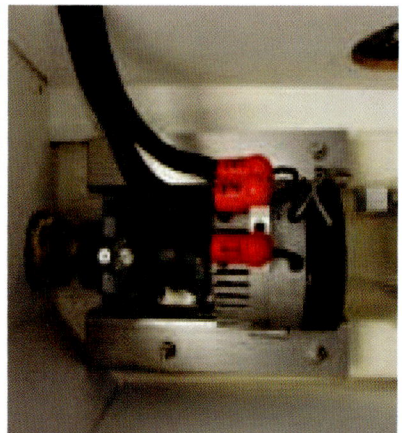

Parallel Hybrid System

A parallel hybrid maintains the mechanical connection between the engine and propeller shaft. As it's name implies the electric motor acts on the drive shaft in parallel with the engine.

You can drive the propeller directly from the engine or from the electric motor or from both. You can also disconnect the propeller for a stand-alone generator function. During re-generation the engine is disconnected.

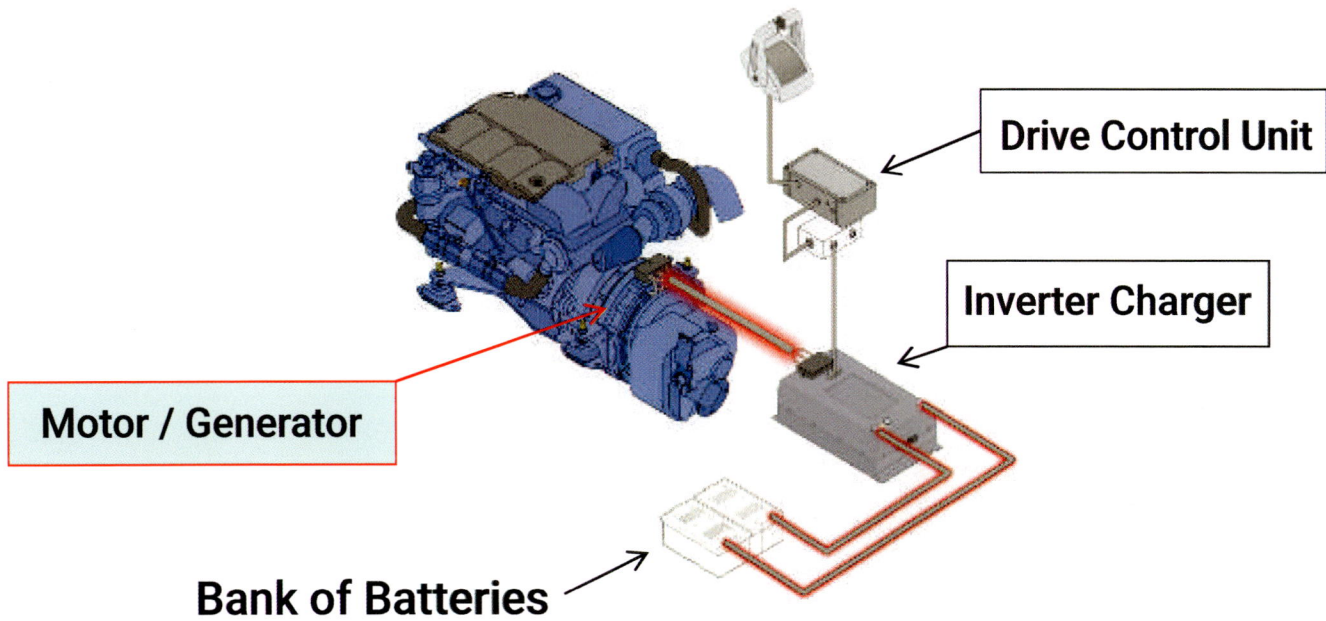

Boat Wiring Schematic

AC Panel- DC Panel - Dashboard and Electronics

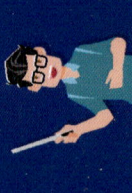

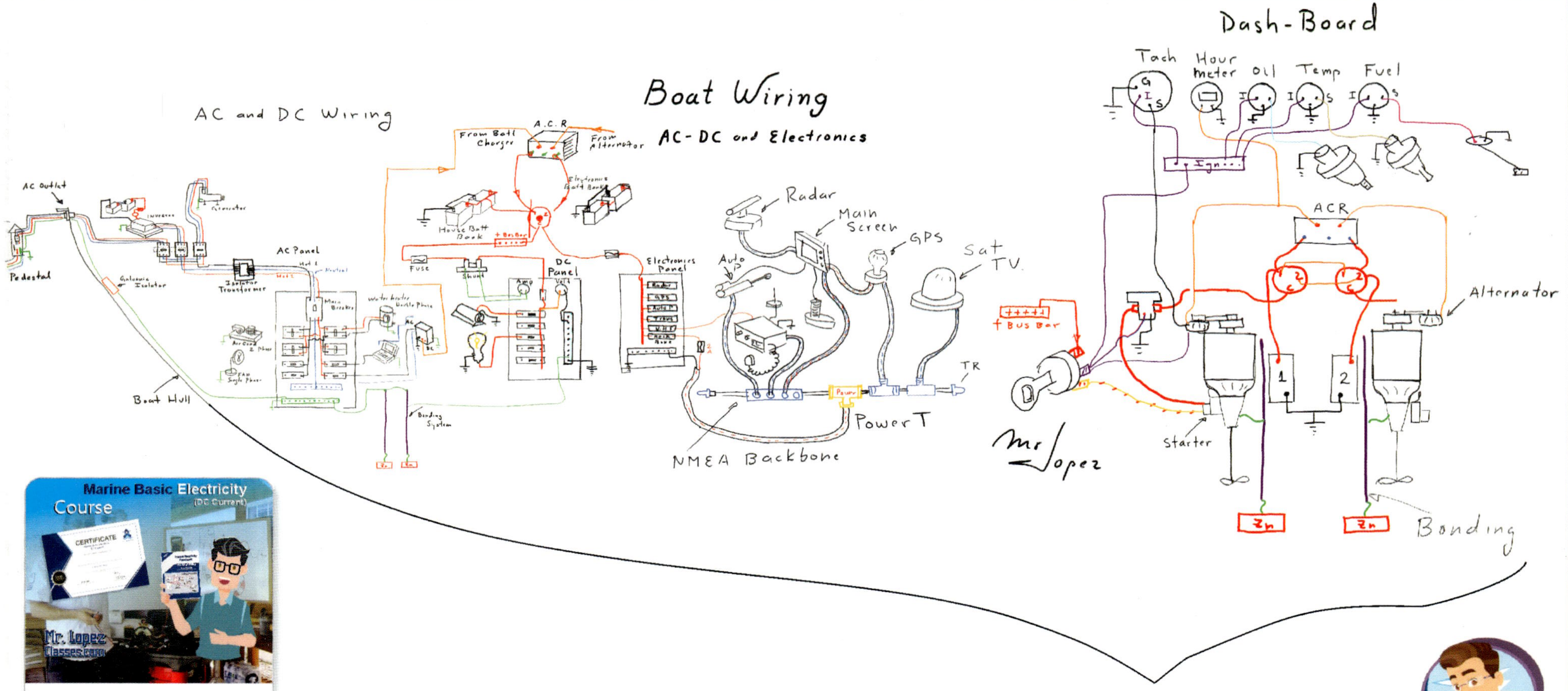